D1648179

The Ice Palace That Melted Away

The Ice Palace That Melted Away

Restoring Civility and Other Lost Virtues to Everyday Life

BILL STUMPF

Pantheon Books New York

Published in the United States by Pantheon Books, a division of Random House, Inc., New York, and simultaneously in Canada by Random House of Canada Limited, Toronto.

Library of Congress Cataloging-in-Publication Date

Stumpf, Bill.
The Ice Palace That Melted Away: Restoring Civility and Other Lost Virtues to Everyday Life / Bill Stumpf.
p. cm.
ISBN 0-375-40221-7
1. Technology and civilization. 2. Quality of life. 3. Human comfort. I. Title.
HM221.78 1998 98-13346
CIP

Random House Web Address: http://www.randomhouse.com

Book design by Mia Risberg

Printed in the United States of America

First Edition

2 4 6 8 9 7 5 3 1

for Sharon

ACKNOWLEDGMENTS

Of all the benefits life can deliver to a person, the ones I value most these days are associated with experiences and friendships. Professional and otherwise. I don't like to think of myself as greedy or selfish, yet when I realize how a few experiences and people have marked my life, a sense of hoarding, coveting, and withholding from others overwhelms me. They have and continue to shape my life and give me hope for the future. What's more, some of these experiences I can only imagine (like riding on the back of a goose in flight) and some of the people who have stimulated my heart and mind I've never met, like Christopher Lasch and A. A. Milne, Tennessee Williams, Duke Ellington, all of whom are dead, but have left me a legacy of words, ideas, and music to feed upon. And there are others I don't expect to meet, like Robert Hughes.

Among those I have had the privilege of knowing, befriending, and directly working with and to whom I owe much gratitude are D. J., Hugh, and especially Max DePree, who believed in my talents more than I did early in my career and literally made investments in my future. And on a workaday level, Clark Malcolm (my friend, BS detector, and editor), Mickey Friedman, Julia Child, Ralph Caplan, Pep Nagelkirk, Gary Miller, Mary Catherine Bateson, Steve Frykholm, Don Chadwick, and Edward Wohl, all of whom have wittingly and in some cases unwittingly become my most cherished gang members, and also my agent, Sandy Dijkstra, who all along had faith in the manuscript.

And of course, Sonny Mehta, Shelley Wanger, and Pantheon Books for publishing this book and sharing my concerns about designing a more civilized form of daily living.

CONTENTS

PART III

Paths of Civility

INTRODUCTION

It is a lack of confidence, more than anything else, that kills a civilization.

—KENNETH CLARK

Things fall apart; the centre cannot hold;
Mere anarchy is loosed upon the world,
The blood-dimmed tide is loosed, and everywhere
The ceremony of innocence is drowned;
The best lack all conviction, while the worst
Are full of passionate intensity.

—W. B. YEATS, "The Second Coming"

Difficult as it is for me, an industrial designer living in the Midwest, to describe ideas in words instead of illustrations or objects, I have started writing this book many, many times. For me, words seem only like a beginning, so I couldn't help sketching out some ideas and observations to go along with the words. I've been lucky in my career, for a good share of my ideas have actually become products. Yet, I feel I've barely scratched the surface.

I've been thinking for thirty years about why we live the way we do and how design hinders or enhances our lives. Civility and grace have been on my mind for a long time. All these years I've been observing a change in American life that the optimist in me (designers have to be optimists) describes not as a decline, but a roller-coaster search for comfort. Not comfort as in "the lush life" or "creature comforts," but comfort as defined by William Gass: "a lack of awareness." If your shoes are truly comfortable, you aren't aware that you have them on. We seem to be running in a marathon today, clomping along in wooden shoes. Those of us who notice are uncomfortable, and a lack of civility seems to be the reason. As an industrial designer terminally preoccupied with the quality of life and human artifacts, maybe I'm merely overwrought and suffering from what designer Jay Doblin called "the curse of aesthetics." But the quality of private life—colored, as always, by our public lives in this fin de siècle world of ours—seems to be suffering. How civilized is our world? We seem unhappy in the face of restrictions on rights or speech or fashion or manners or sophistication. Many of us are even unhappy with the way we live. My own explanation is that we are unhappy because every day we see the chipping away of a civilized way of life all around us. We notice.

I recently listened to a cabdriver on my way to the Loop from O'Hare in Chicago tell me why he emigrated from Johannesburg, South Africa. "I know racism is a big problem here in the States, but I also know America openly talks about its problems and eventually solves them." I couldn't agree more. *The Ice Palace That Melted Away* is my way of talking about some of our problems and ways to remedy them.

I think of civility in three ways—things of civility, places of civility, and paths of civility. Civility is the something extra—the added measure of grace—in the way we shape human behavior through objects and custom. Civility is comfort, hidden goodness, social lubricant, personal worth, helping others, play—civility is the joy we take in our human achievements and the compassion we show toward our all-too-human faults. Civility can be extended by technology and can be obliterated by it. Civility is toleration, understanding. It is the integration of differences, not the heightening of them. Civility can be found anywhere—in the great faux city of Las Vegas and the backwaters of the Midwest. I have seen civility in many places, and I am writing this book to show that civility does exist and adds daily to the potential of human beings to move beyond survival to real achievement.

I wonder if we are living in civilized times. Some cultures may believe they are more civilized than others. Throughout history some places have grown, and some that were once considered to be civilized have deteriorated. Today we see major ebbs and flows of civil behavior in developed and underdeveloped countries. We have witnessed and continue to witness the wrath of racism, violence, hatred, and everyday meanness in our own country—in government, corporations, churches, law enforcement, schools, hospitals, neighborhoods and roadways, and sadly even in the design of toys kids play with. In an ever-shrinking yet culturally diverse world, we need civility to survive.

My views come from observations made over the years in many countries, including the United States. A designer's view, for what it's worth. Another industrial designer, George Nelson, told me long ago, "Study life, not just design." My civic ideals have been shaped by good and bad experiences and by fond memories of my childhood in a densely populated, post-Depression-era, blue-collar urban setting in south St. Louis, Missouri, where I was often left free to explore and roam the neighborhood, where there was much to learn.

A civil society, it seems to me, is balanced, healthy,

and creative, a bulwark against economic expediency, a prime requirement of economic health. My experiences, dreams about a better life, feelings that life is missing something more than modernity and yes, postmodernity, has a mind to deliver—all make me take time to write this book.

No thing is too large or too small to have within it a civil message—inventions, all manner of urban architecture from public schools, day care centers, to housing, police cars and uniforms, taxicabs, food, plumbing, telephones, computers, media, affordable and available products of quality. Levi's jeans say a lot about America; so do jazz and rock music, all of which were and remain trickle-up products. Things of civility come from all over, it turns out.

They particularly come from the brains of individuals willing to take a risk. Emerson claimed that a genius is a person convinced that his or her perceptions are the perceptions of millions. I believe him. Things of civility can be luxurious, frivolous, or plain and simple. There's more than a grain of truth in Frank Lloyd Wright's wry demand to give him "the luxuries of life and the necessities will take care of themselves." Even a bum looks good wearing a fresh flower in his lapel. A white linen tablecloth doth make a pretty meal. Lacy underwear has its charm. Maybe some excess is okay: spending an unlimited amount of time listening to my five-year-old grandson talk about anything he wants to; taking time to polish the silver for Thanksgiving dinner; going into debt to buy a Steinway piano or sail around the Cape of Good Hope. As impractical and unfashionable as excess is today, the suspension of time and money in the pursuit of art and science and community creates its own economy, an economy in which treasures are born, an economy in which civilized life can thrive.

I rejoice in places of civility that promote trust and goodwill. How well we think about ourselves can be seen in the quality of our homes but also, and perhaps more important, in the quality of our public life. I rejoice in seeing families with children in Central Park and neighbors hosting block parties. Places where a kid can leave a bike in the front yard, where it's okay to walk alone at night. Our public places of civility tell us something about what we think of ourselves. Increasingly the image and substance of things public rise from the private sector. Yet who can truly feel at home in the courtyard of an AT&T skyscraper.

I believe civility manifests itself in paths of civility. People like D. J. DePree, the founder of furniture manufacturer Herman Miller, Inc., and poet Maya Angelou blaze trails for the rest of us. People who were or are living examples of setting about being good, not just wise guys, or pundits, or preachers, or tycoons, or saints. People characterized by a sense of grace, with good humor, with style—a kind of attitude and behavior that transcends greed and self-interest. The paths toward civility that they leave us have become increasingly overgrown. It's time to clear them and use them.

Civility is a matter of opinion and often ambiguous. While I can rant about hyper-commercialization and wanton materialism in America, it's hard to keep from smiling when I think of McDonald's golden arches hovering over Tiananmen Square or America Online connecting fellow cyborgs in Belgrade and Beijing. Business, however crass it has become, has to be given some credit for infiltrating hostile and oppressed societies and building relationships beyond our shores. I just wish they'd get around to freeing up Cuba and the Havana cigar market.

I believe the paths of civility are enduring and sustainable. The most destructive feature of modernism was the false assumption that the future had to be wrought out of the rubble of the past and present. Oh, yes, design and designers have had their uncivil moments.

An old Norwegian house painter annually touches up the paint on our house. I had always put off painting the house until it looked shabby, and then the project always amounted to an expensive,

monumental summer travail. This particular painter has taught me a better and less expensive way, a way practiced in Norway and the Netherlands. Mind the paint regularly and touch it up, and you may never have to repaint the whole house again—and the house will look fresh all the time. It's time we all began to touch up the civility of living in America, lest our house rot from under us. Columnist Ellen Goodman has said, "Civil rights protect individuals. Civility protects community. Individuals plead their own cases in courts. Who pleads the case for community?"

As an industrial designer, I am moved to do something about civility, to design a better world, to rig a better existence—to compose a more melodic life, as the anthropologist Catherine Bateson might put it. To save our culture and our world, though, we must all lend a hand. Neither designers nor politicians can do it alone.

So I decided to make a public appeal. Rather, I decided to call my friend Clark Malcolm and ask him if he would help me write this book. After all, I said to him, we have books that purport to tell us how to succeed in business without trying, how to win friends, or lose weight. Why shouldn't we write a book that claims to tell us how to save the world? For better or worse, he said yes.

At least I can tell you how the world is being saved in bits and pieces, here and there, by people and communities that see a problem. Things aren't hopeless, after all. There are people who remember what it's like to be children, people for whom imagination appears daily on their agendas. People do make choices about how they live and what their world looks like and how it works. People use technology without worshipping it, and they preserve the past without living in it. People everywhere haven't forgotten the advantages of being connected to reality and the natural world.

In any case, I'm an optimist about the whole thing, this mess we seem to have gotten ourselves into. I don't know all the answers, but I have some inklings.

PART I

Things of Civility

1

The Design of Flight

The power to throw yourself happily through the sky, to see the familiar world from any angle at all, or not to see it, to turn one's head and spend an hour in the other-world of the hills and plains and cliffs and lakes and meadows all built of cloud. . . . So it is that many people travel by airplane, but few know what it is to fly.

—RICHARD BACH, Biplane

Postmodernism reminds us every day that we live in a "postnatural world." Social philosophers like Walter Truett Anderson preach a sobering yet liberating view that life and how we live it is a matter of our own choosing. The media laments the loss of "civility" and "manners." Everybody seems unhappy when confronted with things that don't work.

I deplore fulminations about impending disaster (though I'm guilty of a few) and sentimental invocations of past innocence. I'm old enough to remember what it's like to have a tooth filled without novocaine. I'm also old enough to remember those valuable ideas and rituals that simply got waylaid or discarded in the sweep of the mindless change and cultural lurching about so typical of American life. I say, If it ain't broke, adapt it for next year—but not in abstract or metaphorical ways like the "lick a brick" postmodern architecture of the 1980s.

It's not that I've given up on achieving a civic agenda, nor do I feel an urgent need to call for a born-again spirit of design. It's simply a matter of time. Our binge of consumption may be running out of steam: real-time experiences become virtual; furniture, homes, and tools become commodities; the marketplace becomes automated and cyber-franchised. We need a new ethic: replace consumption with play; find an enduring and sustainable way of coexisting with the environment and one another; remove the tether of debt; lose the fear of being unfashionable.

If I'm going to talk about design, that purely arbitrary and immensely human construct, I should say that by design I mean the process both physical and mental by which people give an order to objects, community, environments, and behavior. Like many hard-to-define but profoundly important activities, design is both art and science. It aims to make our existence more meaningful, connect us to natural realities, show us the advantages of graceful restraint, infuse serious work with playful humor, extend human capacity—physical and emotional and spiritual. Designers make ideas into things.

But who am I to save civilization. Perhaps I'm a lot like you. In spite of the apparently sacred individualism I enjoy as an American, I honk my horn at a way of life that seems increasingly difficult and temporary and rootless. Like you, I procrastinate and rationalize. I gave my first lecture about civility in 1979 at the Walker Art Center in Minneapolis.

I believe that designers traffic in the saving graces at the heart of civility. A sense of wonder. Goodwill. A second chance. Extra headroom in the car, daylight savings time perhaps, the postman who brings in the newspaper, the extra doughnut in the bag, the airbag in your car, a corporation that recycles materials, prudent research, natural antibodies that fight disease, extra toilet stalls for women at the new baseball park, clean windows on the bus. An apple free of chemicals. A breath of unpolluted fresh air. A safe street. A humane airplane.

I love to fly and travel. But for Pete's sake, why can't more pleasure and excitement and civility be safely integrated into this common experience? Why has flying become such a dreary and dreaded activity? What's missing—better service, food, and more video channels?

The first American astronauts had to persuade

NASA to install a window in the early space capsules. Apparently to space engineers, the visual feast of seeing Mother Earth from space for the first time in human history was an *unnecessary* experience. John Glenn was little more than a laboratory monkey: a biological feedback subsystem in a space capsule. Strict techno-utilitarians still argue that manned space flight has been a waste, that computers and sensory devices can scan the heavens more efficiently than humans. Depends on what you're looking for.

This same sort of economic-technical rationality drives the design of modern jet aircraft. As passengers, we are reduced to little more than eggs in a carton, packaged safely (and, of course, as efficiently as possible), destination-bound with no expectations. We are fortunate to fly in relative safety, even though many would say our air travel system is in decline. Like most of technology, we depend upon it but have come to take it for granted. I still find it utterly amazing to circumnavigate the Earth with such little effort and so little pleasure. But there is nothing *normal* about traveling at 600 mph at 35,000 feet in subzero temperatures, in an oxygen void, in a largely handcrafted structure composed of millions of pieces, alongside tons of explosive fuel, and guided by instruments.

It's astonishing to see the giant of them all—the

Boeing 747, "fat walrus with wings"—move, much less fly. I love the bulbous presence of the Boeing 747, on the ground or aloft. Especially at dusk, before an intercontinental flight, peering into the eagle-eye windows which reveal the cockpit ablaze with a miniature cityscape of sparkling lights, I envy the pilots cocooned in their cozy den. What a toy! Like locomotive engineers or steamship captains, they command one of the greatest Earth-encircling machines.

When I step into the 747's cavernous fuselage, I begin to exchange thoughts of where I am with images of the distant place I'm headed. Buckled into my padded, grossly uncomfortable seat, I await the whine of the engines and the coming rumble, roar, and thrust into the sky. It can stay aloft for endless hours while carrying the population of a small town. It is the great ethnic pollinator of the twentieth century. But that's where the pleasure stops.

In the stale air of the cabin, bored cabin minions dispense prison-quality food; on come the commercials; out come the laptops and, if one's lucky, sleep sets in until the landing. I'd prefer suspended animation. On one such flight, I watched the movie *Memphis Belle*, a thrilling epic of flying in the venerable World War II bomber the B-17, also a Boeing design. While in no way would I or anyone else in their right mind fantasize about flying dangerous missions over flak-filled enemy skies, I envied the pilots and crew flying in a plane which seemed closer to the nature of flight. How lucky those nose, belly, and tail gunners were to sit in their Plexy bulges with 180-degree views of the earth and sky. Talk about childhood dreams come true!

As I mourned the loss of wonder in flying and asked myself whether it could be redesigned into modern aircraft, I started to doodle over a printed photo of a 747 I found in a magazine. Imposing on its fuselage some of the Plexy bulges, nose windows, tail windows and such found on the B-17, I dreamed on. Flying would seem less dangerous to me if I knew more about the flight plan, the speed, the navigation, the fuel reserves, other planes nearby and, more importantly, the glorious skyscape I was traversing. (Since I began this book, some airlines have started showing this sort of information to passengers in flight. Let's hope it becomes routine.) Imagine the glory of watching the sunset from the nose or leading edge of a wing. Imagine a windowed perch on top of the tail where one could view the 747's flexing wings and shimmering architecture, punching holes through the clouds, spewing vapor trails in the moonlight. Modern fighter planes have sophisticated night vision and ground telescopes that reveal small animals from 35,000 feet, so why can't this technology be plumbed to every seat in a passenger plane? Imagine your own telescope scanning the stars and the moon from the clarity of the night's sky. Instead of watching boring movies, a faux cockpit could be virtually realized on a personal screen, with interactive controls. This would make us a part of the wondrous technology we have created.

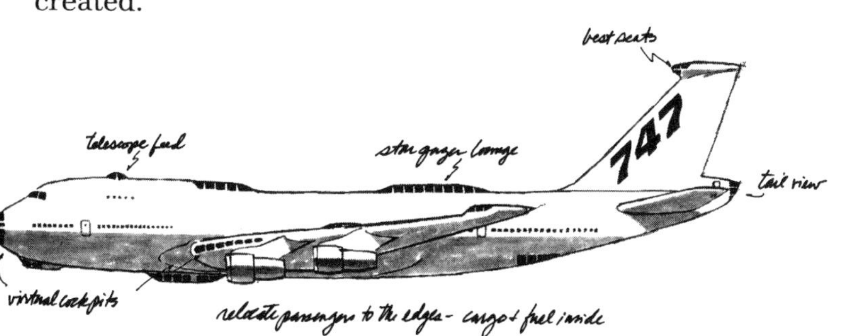

In spite of the million-plus air miles I've flown over the years, I realized I had never experienced the thrill of flying. And I wondered to what degree in civilized societies the thrill of an experience is inseparable from the experience itself. Johan Huizinga, the Dutch historian, cautions us, in his discussion of the Middle Ages, not to forget how much we have sanitized natural experiences. We are warm in winter, cool in summer; we light up the night and darken the day. Our technology has allowed us to separate the visceral reactions from the experience producing those reactions. Maybe it's time to change that. Maybe civility can reconnect the two.

To test out my theory, on a warm, sunny fall day, I chartered an open cockpit Stearman Biplane, a World War II trainer, for an afternoon of flying from a local field in Minnesota. Albeit a passive observer—I'm not a pilot—I spent a couple of glorious hours in Saint-Exupérian ecstasy strapped in tightly, skimming the landscape of the upper Mississippi Valley. We did sharp turns, dives, and thrilling side rolls over a few oblivious cows. I felt the G-forces in turns and coming out of dives and the exhilaration of weightlessness, and was frankly scared a few times that I'd fall out of the plane's harness. I tried the stick and found that planes don't track like cars. It was sheer unearthbound fun. Wearing a soft helmet and headphones, I eavesdropped on radio chatter from the big regional airport. Conquering my fears, I exhorted the pilot, a seasoned Northwest Airlines captain—who flies these planes for the same reason I wanted to: to get closer to the nature of flying—to give me my money's worth. He did, and more. Upon landing, a sweet taste of hot engine oil was on my lips and face, a natural spray from the old radial engine uncowled in the nose.

Can't experience and emotion be wedded forever? Surely design can reunite experience and emotion, a wonderful and all-but-forgotten part of a civilized life. Like the vertu of the Romans, the Rule of the

Benedictines, the enlightened humanism of seventeenth- and eighteenth-century Europe, the written and unwritten codes of the Victorians, design can become a civilizing force in our lives.

2

My Dad at the Brewery

In most eras voices cry out against the visible decadence; for every generation—and especially for the aging—the world is going to the dogs.

—JACQUES BARZUN, The Culture We Deserve

With easy access to forms of technology our grandparents never dreamed of, we long for more humane work, a "kinder, gentler" society, classicism in architecture, comfort in our homes, user-friendly products, and a sense that as public beings we deserve more. We are drowning in technology, most of which exists only to make us momentarily more comfortable or to relieve us of the duties of survival that engrossed so many of our ancestors. Yet we try awfully hard to deny this fact. Americans in particular are a comfort-obsessed society.

My colleagues in industrial design and I try to make people comfortable—aware, yet unaware, at the same time that a designed object is making their lives better. We design shoes or clothes or chairs or streets or bridges or public toilets. All have one goal in common—comfort. When they are aesthetically pleasing and functionally correct, we tend to be unaware of them.

My ideal for us is a sublime, simple state of understanding and wisdom—an unself-conscious state of comfort—where our stomachs are full and we are breathing clean air. We need to retreat from our hyper-awareness of ourselves, our world, and the hideous shortcomings of modern life—including the disappearance of simple civility. But we inexplicably retreat from the wrong things, welcoming apathy in the face of horror. As journalist Christopher Hitchens said in a speech given at the IDCA Design Conference in the spring of 1993 in Aspen, Colorado, "Although I read about the Holocaust, and I've lived in its shadows, upon visiting Bosnia I can't believe I'm bearing witness to a world willing to stand by and do nothing about ethnic cleansing."

Who hasn't seen the homeless adrift on the street, been approached by a beggar, had our windshield cleaned for a tip at a stop sign, and simply denied the experience? In 1996 over 170,000 women were raped in the United States, and thus we are denying women the most simple freedom to walk safely down a public street. If men were the victims, I doubt that this abridgement of a basic liberty would exist. We are cocooning ourselves in our suburbs, with our home electronic security devices, our hermetically sealed office buildings, our gated communities without a thought for what happens when we fly out into a larger world.

In the late 1940s, ignorant of today's menu of problems, I enjoyed a blissful unawareness as an adolescent in St. Louis. Not only did I live in a house without working locks, I roamed the city alone on foot and on streetcars unsurveilled by parents or police. I grew up in a lower-middle-class neighborhood with no reason to fear anything and with good reason to think that one day I could do something that would have an effect on my world. I sorely miss that feeling.

Today social critics and media pundits lament and whine about the decline of Western civilization. Yet no one has devised as useful an antidote to misery as Jelly Roll Morton, who somehow transformed the brothels of New Orleans into music and community. If he could succeed with such unpromising material, surely we may yet build community in our worlds.

My dad was a machinist who worked for Anheuser-Busch in south St. Louis, one of the world's biggest breweries. He died when I was thirteen. Of all the activities he shared with me, including ball games and hunting, by far the most wonderful was a trip with him to work. I was enthralled by the smells, the noises, the vast factory spaces, the robotlike motions of giant machines fill-ing bottles at lighting speed in the beer-making process. Growing up, I never looked on my dad's toiling in a brewery as a less than noble vocation. Like great steam locomotive engineers or pressmen in a printing shop, my dad controlled enormous machines. That was something. Working the controls of a huge pasteurizer was to me every bit as impressive as wielding a scalpel or pushing a pencil. There was plenty of dignity in work.

Today children and adults seldom see what goes on in factories, except on an occasional field trip. In America, the manufacturing of everything from automobiles to breakfast cereal to hamburgers goes unseen. Like videocassette recorders and portable radios, factories have become mysterious black boxes that we have no business opening. They have been relegated to industrial parks suspected as sources of pollution. The most modern mass-production facility in America, the Georgetown, Kentucky, Toyota plant, lies hidden from public view in the hinterlands of horse country. Even a drive past these plants reveals nothing of their purposes. No smoking stacks, no noises, no smells—they stand as mute monuments to modern production. It's almost as though we think something sinister is going on here or that factory work is bad form.

Only farm kids have the luxury of learning how things grow, how animals are born, how crops are harvested—*how life actually works.* Most suburban children associate work with washing dishes or cutting the lawn. Industrial technology for many of us is expressed in the workings of the local car wash. This, of course, is nothing new. The world has been thus for a long time, but there is something more insidious than simple ignorance going on here.

As a culture we seem to be increasingly indifferent to common labor. Hide it, put it away, let somebody else do it, send it overseas. Rosie the Riveter was someone's sister, wife, or mother in World War II. Today our heroes work in offices and business suits. Is this why we're indifferent to poorly paid people in sweatshops in Southeast Asia who make the sneakers we wear as we shop? We love to consume but have developed a squeamishness about the processes that make what we buy.

George Nelson once said, "American cities are by and large utilitarian and not cultural centers like Rome, London, or Paris." He's right. There was a time not so long ago when the very mention of a place like Chicago meant hog butcher, and San Jose meant Silicon Valley, and Milwaukee meant beer. Cities like Pittsburgh, once proud to be known as the Iron City, now profess cultural diversity as their commerce. American cities are striving hard to shed their exclusively utilitarian images in favor of broad cultural centers complete with art museums, symphonies, and major-league sports.

The image of smoking chimneys, hammering forges, screaming factory whistles, polluted rivers, and heaps of trash has permanently disenfranchised the means of production. Thank goodness some of the more frightening and dangerous marks of an industrialized world have been mitigated in the name of the environment and human health. But what else have we thrown out with the polluted bathwater?

In my hometown of Minneapolis, the populace went crazy at the sight of clean steam clouds emanating from a state-of-the-art city incinerator plant near the downtown area. Now this kind of sensitivity is admirable when it is directed at some local polluter, but let's face it, factories are the engines of growth in a consumer economy. We live by the rules of universal consumption, yet we increasingly deny its place in our own backyard. The manufacture of a Boeing 747, or Ford Taurus, or Apple Macintosh is as much a part of a civilized world as tea and crumpets.

Imagine that McDonald's restaurants, the world's largest retailer of hamburgers, were redesigned to honor labor and production as well as merchandising and speed. What would it look like? Imagine that they baked their own buns on the spot. How much better it would smell! Imagine machines making ketchup, mashing tomatoes, and spouting steam. Or machines grinding meat and pressing it into burgers. Perhaps the economies of scale necessary for McDonald's to sell us their food so cheaply would preclude a bakery in every franchise.

But what would happen if we could witness the activity and energy of the staff making our meals—in the heart of the restaurant in full view of you and me. A restaurant I know of treats its customers to a regular show of chefs and assistants doing a culinary dance behind large glass dividers and the food issuing from the visible kitchen. They aren't afraid to show anything. A giant material muncher could recycle all the food packaging on the spot and spit out new cups, bags, and trays. We could give children a sense of production that seemed a part of their lives.

Rarely in America is labor associated with fun. Yet many adults will tell you that they are happiest

and most productive when work and fun commingle. But at McDonald's and every other source of goods, entertainment is divorced from labor. How terrific it would be to be able to take the family along to the automobile plant and watch the family's new car being made from start to finish. We used to be able to tour the great cereal plants in Battle Creek, and thousands of schoolchildren and adults did so every year. No more. The only true labor we are allowed to witness is the quaint hand manufacturing in historical villages and towns. Labor was and is a part of life that we need to learn about and understand.

A civilized society comes to grips with all parts of life, the tragedies and the successes. Let's be honest—life is sometimes dirty. Sure, clean it up, but don't ignore the dirt. People without knowledge are powerless, and at the root of an uncivil society lie Ignorance and Powerlessness and their offspring, Apathy.

3

Bags and Other Small Graces

Oh, if I were doing nothing only out of laziness. Lord, how I'd respect myself then. Respect myself precisely because I'd at least be capable of having laziness in me; there would be at least one, as it were, positive quality, which I myself could be sure of.

—FYODOR DOSTOYEVSKY,
Notes from the Underground

The phrase "pester power," which recently emerged from the British press, aptly describes the daily but hard-to-pin-down annoyances that afflict Americans in myriad ways every day. If you have survived war, urban blight, or human violence, you might dismiss pesterings as a petty, middle-class issue. Intrusions like the sound of jackhammers, jet aircraft, television commercials, junk mail, and unsolicited offers of insurance and aluminum siding do not threaten our lives. Or do they? We aren't obligated to submit to the "tyranny of small things," as one writer has put it. Nuisances can evoke deep feelings of rage in all of us as we bridle at our own powerlessness to achieve anything like serenity in our daily lives. At the very least, design can rid us of such nuisances, can eliminate the constant reminders that our civilization is not what it's cracked up to be.

At a local cinema recently, a man in the audience, incensed by the American Express and Coca-Cola advertisements preceding the movie, booed loudly and persistently. Enraged by the frequent visits of door-to-door salespeople, a man in Santa Barbara altered the homey Spanish saying on his lintel to read "Mi casa *no* es su casa." Certainly we are free to boycott movie theaters and bar our front doors, but shouldn't we rather demand that movie theaters rid themselves of cramped, modular rooms with poor sight lines, uncomfortable seating, TV-format movie screens, marginal sound systems, stale popcorn, and forlorn locations? This is a problem multiplied ten times over in all areas of our society. It is also a problem for design. The architecture of the movie houses should change, not necessarily reverting to the ornate and inspiring palaces of the thirties, but at least to become pavilions of moviedom and folklore. Movie theaters could embody the fantasy, myth, story, laughter, and adventure of films themselves. Outside, towers, neon lights, animated billboards, preview screens—why not? Inside, a transition space would acclimate the eyes to the darkness. The screening rooms would attend to ergonomic principles in sight lines, seating, and air-conditioning. An electronic system for consumer ratings of films would immediately make available to the public the audience's opinion. A balcony would become a cozy place for ardent teenagers and privacy-seeking adults.

Some intrusions are unbelievably wasteful: every day Americans receive enough junk mail to heat 250,000 houses. The morning paper or evening news show or random mayhem on the street amounts to a daily psychological irritant. All these pesterings may add up to something that is equally dangerous. They may even be numbing us to the real tragedies. The building of yet another mall, the death of yet another family farm or small business, the razing of yet another fine old building, the pronouncement of yet another silly and meaningless "Have a nice day!"—these pesterings add up to one uncivilized *un-nice* day. These are small things compared to the violence in the Balkans or racism in the United States. Sometimes, there is something that passes unnoticed with a measure of grace. The ramped curb comes to mind, undoubtedly a vital feature of mobility for older or disabled people, but taken for granted by in-line skaters, bicyclists, and the rest of us. My local post office stays open fifteen minutes past the legal closing time—someone's supplying a last-minute grace period without getting paid to do so.

The offering up of "little civilities" can be particularly unwelcome. The ever present and largely ignored buzzerphrenalia attached to burglar alarms in homes and cars, appliance timers, beepers, and other devices designed to forewarn can be infuriating—or the "armored" packaging of those airline peanut bags and aspirin bottle caps that require Houdini-like genius and strength to open. All of these poorly designed and poorly thought-out annoyances edge us toward an uncivil life.

It's also common for us to look for heroic measures of goodness and grace when and where we should know better, like expecting stealthlike but constant attention from a waiter or an astounding gourmet main course at a fancy restaurant (those with atlas-size menus) when the soup and bread are just average. And it's easy to fall into a resignation syndrome after repeated disappointments. For example, Americans visiting Italy come to rely on the English word *normally:* normally the mail is delivered, but due to a strike, not today. Normally the elevator works, but not today. With such casual resignation down pat after living in Rome for ten years, a good friend, a seasoned traveler, a wine lover, and gourmand extraordinaire, says, "I don't get upset about American wine lists anymore. Bad wine is better than no wine." American television executives boast of the power of marketing and advertising with the hidden help of the unwritten "Law of Diminished Expectations in Mass Media."

While we can be delighted by small measures of grace, we can also take for granted their value, and resign ourselves to the rudeness of life without them. Take, for example, the common shopping bag, in its own right a wonderful design—simple, cheap, made of recycled kraft paper, reusable for a host of practical needs. It is a small but special act of civility on the part of those supermarkets that supply strong paper bags with handles. Handleless bags are terrible. Handled plastic bags are equally bad, for they can't stand up on their own. That's why the checkout counters have those wire hanger devices, but of course you and I don't have the same devices in our cars, or at the front door, or on the kitchen countertop. These bags disgorge their contents without warning, and the desire to set them down out of weariness is muted by their structural infidelity.

Paper bags with handles, perhaps costing a few cents more, allow heavy loads of groceries to be balanced in each hand, allow one to open a door without setting down the load, and are particularly useful for older people with weakened grip strength, a demographic segment I'm fast approaching. Ironically, it's the molded-in handles of plastic bags that have brought back paper handled bags. I collect these paper bags, fold them neatly, stuff them away, reuse them to carry old newspapers to the curb for recycling, but, alas, I feel embarrassed to take them back to the store. I guess we Americans are suspicious of the unsanitariness of used bags. As A. A. Milne observed about snap-on neckties, toting a much-used bag advertises one's poverty.

Being seen with a plain paper or plastic bag has come to say less about a person than carrying one emblazoned with an upscale logo from the Museum of Modern Art or Nieman Marcus or even Target—any brand is better than no brand. We are willingly, even willfully, branded: bags in hand, labels on our clothing, we have become walking commercials. This form of free advertising and its conspicuous application to everything from sneakers to underwear to jewelry demonstrate our meek acceptance of the commercialization of life. Only booze and pornography come in a plain brown wrapper today. Cocaine comes in a clear plastic bag.

I'll cast my vote for plain paper bags—or *sacks*, as they say in the South. For one thing, they give me privacy; they mask their contents. The clear variety exposes everything, even the unsavory. Strangely in my neighborhood, there is a positive status associated with carrying dog droppings around in plastic bags. It broadcasts the social conscientiousness of the dog's owner and has become part of the fashion paraphernalia of walking the dog. With one's pooch in tow, it's to risk suspicion to be seen without a plastic bag in hand. I find it hilarious to stop and talk on the street to a friend while he or she tries to hide the evil, canine-correct bag behind them. In this case, plastic—even clear plastic—advertises one's virtue.

The Swiss, naturally, have designed a better system. Not only have they deployed (every hundred yards or so) on their busy streets and in their parks aerated dump baskets and dispensers of free excrement bags, they have made the bags of opaque green plastic (chlorophyll impregnated), covered with instructions as to their proper use. But I won't go into that here. In this case I must admit the plastic bag has some virtue.

But plastic bags do nothing for alcoholics. Imagine the drunk hoisting his bottle without the sinister anonymity afforded by the paper bag. Or pity the child desperately trying to hide a candy hoard in front of siblings in a clear plastic bag.

It's a shame. I like the anonymity of the simple brown paper bag. Sometimes, I don't want my wife to know I've been to Sam Goody's or the local bakery. Somehow after she shops at Nordstrom's, she never leaves her glitzy bags around the house. And why can't I buy a golf shirt without a stupid little horse embroidered on it?

Maybe small acts of grace matter, after all. Maybe, instead of slowly destroying ourselves by submitting dumbly to pesterings, we can save ourselves with small touches of grace and civility and good design.

4

A Handsome Cab

Perhaps through a general reluctance to be anything but English, the old guard establishment in England has always been more aloof from fashion than the traditionalists in other Western countries. Young Brits, on the other hand, recently have been leading the world of fashion. Modernism touched England but to a much smaller degree than America. Even today, the architecture of Sir Norman Foster's Lloyd's of London building gains much of its drama from its Soanesian context in the financial district and the imposing dome of Christopher Wren's St. Paul's Cathedral. Lloyd's building in downtown Houston, Texas, would merely be more of the same, a variation on the theme rather than a new melody.

In spite of the promise of the cyberworld, I take some glee from the cultural stubbornness and immutable spirit for daily living I see in the British

Isles. After all, it was here that the delivery of fresh milk and mail remained a matter of course in the middle of the blitz. The Amish in the United States seem to possess the same spirit of cultural stubbornness, that is, a determination to weigh what appears to be a replacement against what already works. I endorse their skepticism when it comes to innovation. For years, the Amish refused to install lightning rods on their farm buildings because they believed lightning to be a matter for God to determine, regardless of man-made metal rods. They also recognized the rebuilding of a dwelling by their brethren to be an act that rekindles the sense of family and communities.

All of which brings me to an enduring thing, an expression some might say of Victorianism that has withstood the lurching about of progress—the classic British taxicab. It ranks in importance on travel posters with the British double-decker bus, the Bobby, and the Union Jack. This pressed metal, toylike vehicle with its faux resemblance to classic Bentley coaches steadfastly persists as an icon of British urbanity. This technological throwback, awkward and funny-looking, throbbing behind its diesel-powered engine and distinguished by its nonaerodynamic body, stands tall in London's congested streets like a silk top-hatted gentleman. Civil to the core.

More chauffeured than simply driven by convivial, well-mannered, street-smart urban pathfinders, these cabs comfort the old and the young, the empty-handed and the package-laden. The British cabdriver, who must pass a test in The Knowledge, shorthand for an intimate knowledge of London's streets, seems to find honor in the work of driving a cab in its formal cloak of glistening black lacquer, sparkling chrome trim, impeccably clean windows, and spotless interior. The British taxi and its driver say to me that there can be a haven in a heartless urban world.

The British taxicab experience speaks of first-class public service for all kinds of urbanites. Compare it to its American counterpart: a lurching, rattletrap affair painted school bus yellow, with dirty upholstery, road-grime-splattered windows, permeated with noxious odors from assorted air fresheners, and driven by indifferent driver-rogues, blissfully ignorant of destinations beyond convention centers, airports, and hotels. One has to wonder why we do not ask for more. Even the most selfish, individualistic American would be better served by a more energetic sense of public worth.

To find such levels of grace in American public transportation, we have to search for it, even pay a high price for it in special limousine service or dolled-up rides in the horse carriages around New York's Central Park, rides rich in nostalgia but painful reminders of the joyless state of current urban transportation. We either deny or have forsaken the inherent joys to be found in travel. As Paul Fussell says, "We, at best, tour or commute but rarely travel in the classic sense today." Travel reduced to destinations is merely commuting. We no longer experience flying, only something called jet travel—often in planes appropriately called Air-Bus—which means being imprisoned in a tube like gerbils.

Knowing the futility of integrating playfulness into the efficiency agenda of public transportation, I offer some ideas here that assuage my longings for a different reality within the limits of known technology. Let's call it civil design. To use the taxi as an example, I would take for granted that a new design be safe, fuel-efficient, environmentally friendly, and ergonomically correct. Beyond that, I propose that it have a new spirit, one that celebrates public man, the community, and mutual, practical assistance or service. I would include something of the elegant spirit of the horseless open carriage, not in terms of form but in terms of playfulness. My design for a twenty-first-century taxicab would take upon itself the open characteristics of the traditional carriage

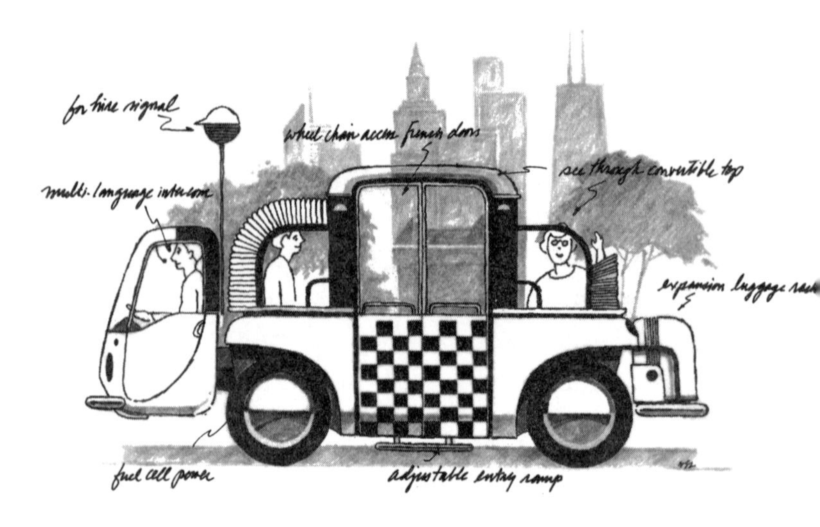

with a glass roof and an occupant-operable convertible top in order to experience the vertical architectural glories of New York or the scent of spring in the air down Chicago's Lake Shore Drive. It would have seats facing front and back, so that conversations could be encouraged. Like its British counterpart it would be more vertical than low slung, jaunty in form, with French doors and a lift ramp and interior tie-downs for wheelchairs. Of course it would be non-polluting and silent running (electric) and have the very latest electronic mapping and destination-finding gear, a multilingual driver/passenger computer-aided communication system, conversation privacy control, and perhaps a bud vase or two. Should the stereotypical New York cab-driver's manners be improved? Nah, that's part of the fun.

And yes, I would include the iconography of good old American taxicabs, the bright yellow color, the checkerboard, and some advertising to boot.

It is possible to combine the old and the new, to merge a civil spirit with technological sophistication—to have fun on the way to a business meeting.

5

An American Palace That Melted Away

The world is too much with us; late and soon,
Getting and spending, we lay waste our powers
—WILLIAM WORDSWORTH, Sonnet XIV

And though our life in this manifestation often turns out to be a bit of trash, still it is life and not just the extraction of a square root. I, for example, quite naturally want to live so as to satisfy my whole capacity for living, and not so as to satisfy just my reasoning capacity alone, which is some twentieth part of my whole capacity for living.
—DOSTOYEVSKY, Notes From the Underground

Christopher Lasch claims that "Rationality has removed the play element from work." The same rationality has also contaminated many of the most common arts of daily living. While I had the freedom to loiter on my walk to and from grade school, children today are programmed every step of the way, from school buses to classrooms and back.

Play, according to Johan Huizinga, is "the pursuit of gratuitous difficulty." Can the longest distance between two points be reconciled with modern capitalism in a digitized world? We have too many people searching for the *shortest* distance. Certainly shopping in a consumption-based economy should be an enjoyable, enriching experience. Yet even here, play has been rationalized. At the world's biggest shopping mall, the Mall of America in Bloomington, Minnesota, play or any form of gratuitous difficulty in the exchange of goods and services has been replaced with computer-driven efficiency. This is no place to have a prolonged conversation with a salesclerk over the qualities of the merchandise. This is no place to exchange pleasantries with a beggar or street musician. This is no place to strut along the street in your finest attire. This is no place to lurk in dark passageways or transact a pound of flesh. There is no time for connoisseurism at the Mall of America, of finding a shopkeeper who knows everything about cigars, or golf clubs, or cosmetics, or shoes, or books. Here is a triumph of stereotypes, a monocultural and antiseptic boredom. Serendipity need not apply.

A friend of mine once bought a camera in a small

town in Ireland. After a leisurely explanation of the options, the proprietor of the camera shop took out the box with a camera inside: "What a nice package we have!" said he. The owner proceeded to put the film and battery in the camera, explain all the functions, fill out (and offer to mail) the warranty card, and dispose of the empty box, all the while keeping up a conversation with my friend about Ireland. This was not simply a business transaction—it was a real purchase!

Oh, for the multicultural splendor of an Eastern Market in Detroit or an ecumenical celebration of Christmas. Oh, to see the more casual perspective on time from other cultures thrown more generously into the stew of life in America. Whoever said the United States has stopped melting life in its pot? Whether we think of American culture as a salad or a mosaic, I think we may have gone beyond melting to meltdown.

People are prepared to go to great lengths in search of grown-up play. In 1992, the St. Paul Winter Carnival commissioned a grand palace to be built entirely of ice. An architectural competition was held, and a number of local firms offered up designs pro bono. The winning entry by Rust Architects was built in a park just north of downtown St. Paul. With donated architectural and engineering services and retired "lake ice harvesters," who worked endless hours in subfreezing temperatures for free, the palace was completed for the carnival's commencement.

A stunning castle made up of 350-pound blocks of ice soared into the sky. Towers of glimmering ice cemented together by slush reached more than a hundred feet over a fanciful plinth set in a snow-white glen overlooking a lake. Civil engineers devised a system of cutting the ice in great blocks and sliding them to cranes. Over a hundred workers toiled into the night for two weeks. Electricians crawled up inside the edifice to string colored lights. All the

St. Paul Ice Palace

labor was donated. Thousands of bundled-up children, adults, and out-of-town folks gathered daily in the frigid weather to watch the palace rise. Over a million and a half people visited the castle, and I was one of them, along with my grandchildren. At night, the candy-colored, semitransparent palace came to life, set against Minnesota's ink-black winter skies, a joyful monument to sheer public pleasure, a work of public love.

In a country preoccupied with speed, utility, and function, here was a grand edifice that would soon melt away. Nonsense, you might say, to spend this kind of effort when people are unemployed and homeless. Not so, for this kind of enterprise is a matter of the play spirit of humankind. It was a grand example of our need for play. Like making a birthday cake from scratch, or trimming the Christmas tree, or wrapping a present, or primping for a date, or polishing the Thanksgiving silver, building this ice palace enlisted great effort and resulted in sheer pleasure, not economics. It is the pleasure found in true work, not the drudgery of mere labor.

Not much of our built environment and not many of the objects we use reflect the innate need for people to play, to pursue gratuitous difficulty. Imagine the experience of riding the Staten Island ferry to work in Manhattan, of standing near the bow with the glory of the New York skyline rising out of the bay. Or imagine a new kind of airplane, one with windows on the wings' leading edges so that a traveler flying west could watch the sunset over the Rocky Mountains. Efficiency in flight may not be everything after all. Cocooned in electronic, automated cars on programmed freeways—what a cold vision of future travel! This highway planner's dream ignores our need to take the top down on a nice day, or feel the road, or maybe get lost on a Blue Highway.

People are fighting back. Rebellion against monolithic, sterile utility is in the wind. Bucking the trend toward enclosing public space, new baseball parks like Baltimore's Camden Yards and Chicago's Comiskey Park are not simply about purist baseball or nostalgia. They are a reflection of the fact that exposure to the elements is worthwhile. They put us in touch with the real world. The designers of serendipity are at work here.

Play and work for the love of it are parts of civilization. They add immeasurably and immensely to our experience of life, to our connections to reality and each other. When the St. Paul ice palace had melted away and the giant blocks returned to water (natural recycling), the most important quality remained: a sense that people are part of something bigger. A community of civilized human beings connected to one another existed here. That feeling still remains.

Maybe getting there is as important as being there. Has the idea that the best trip is the shortest distance between two points run its banal course in American life? Probably not.

6

Just One for Life

Nobody thinks that a self-tied tie matters; nobody is really proud of being able to make a cravat out of a length of silk. I suppose it is simply the fact that a made-up tie saves time which condemns it; the safety razor was nearly condemned for a like reason.

—A. A. MILNE, Not That It Matters

In the literate, rapidly changing, interdependent world of today, whole populations may find their moods oscillating between hope and despair or temporarily absorbed in some overwhelming obsession or fad.

—MARGARET MEAD, World Enough

Some social observers see a decline in the level of civility as merely a change entailed by an increasingly diverse society. The rest of us see it as the source of great misery and discomfort: a lack of those qualities which make daily living possible. Added to this is the American penchant for a never-ending series of panics. Our attention span seems to last all of an historical five minutes. In the sixties, environmental apocalyptic panics made subterranean housing fashionable. Now, it's all but forgotten. The oil crisis made GM downsize its cars and eliminate frivolous convertibles; the company reversed itself in the 1980s. Fear of antibiotics, hormones, and dioxins led to the demise of factory-raised chicken in favor of free-range poultry and the reinvention of milk in glass bottles, now an "in" thing on the East Coast.

I have watched my grandchildren go through a proto-consumer toy acquisition phase of their lives. Want, want, want. Of course my wife and I, doting grandparents, buy them toy after toy, much to their parents' chagrin. Eventually, they lose the rabid desire for more and more toys and begin to discriminate. My writer friend and collaborator on this book tells me that most inexperienced writers, thrilled with language and words, load their prose with every large word within reach. They usually get over this and begin to search for the plainest words. Once in a conversation with Julia Child, I asked her about new cooks and new pots and pans. She admitted to the charm of all kinds of utensils, but now she likes old things that work. The late theologian Reinhold Niehbur writes that "in the end, consumption is not a heroic enough ideal" to sustain a culture. Sooner or later we will wake up. We have, after all, been on a shopping spree in the United States since the end of World War II.

Does design help set off cultural spasms in a market economy? I will say yes, because I have struggled periodically as a designer with the efficacy of a so-called new design, or asked about the perceived and real needs for this or that, or argued with a marketeer about an arbitrary change in an otherwise good design. Through their willing participation in the pursuit of what's hot and what's not, designers assist in the onslaught of unnecessary change—some even thrive on it. "Never leave well enough alone," said Raymond Loewy, pioneering industrial designer and father of three-dimensional advertising (the Coke bottle).

Without a doubt, the best consumer is one who is insecure about his or her place in the scheme of things—particularly children. Fashion, the cut of one's clothes, the style of a haircut, the status of home or automobile, the chicness of shopping places and eateries, the choice of public or private school, generational affiliation (I myself am post-Depression and pre-Boomer), the struggle to be politically correct or incorrect—every decade seems to bring with it the need to redefine the world. The engine of consumption would soon stop without frivolous change. Oddly some of the dons of fashion, the

Calvin Kleins and Armanis, dress in almost priestly garb, simple T-shirts, black-on-black trousers and shirts, while annually foisting hemline revolutions on the public.

Now the phenomenon of the computer, the electronic office, and electronic shopping are threatening to reduce (if that's possible) our lives from couch potatoes to larvae ripening in electronic cocoons. We effortlessly drink in nourishment for our unstable egos; we can't discover who we are or what we want; we are withdrawing into the false security of our painstakingly decorated home-nests. I wonder why we take so seriously what in the long term proves to be folly.

I have vivid memories of my strict, Swiss-born, maternal grandfather's efficient and ritualistic personal living habits. Somehow he created more from life with less. I don't remember him with special warmth or his times with nostalgia. I do remember his way of living as a sign of personal restraint and a seeking after the substance of life and not its appurtenances—a sense he had that life in and of itself was enough.

My grandparents were lower middle class, lived in an unair-conditioned, three-room flat in a densely populated neighborhood in south St. Louis. They knew everybody on the block, had no car, and created a productive garden in their tiny, 1,000-square-foot backyard. Every day, they walked to the farmers' markets, and on Sunday afternoon they had barbecues at family-oriented taverns. They always bought fresh bread at ethnic bakeries on the corner, had fresh milk and produce delivered to their doors, felt safe in their beds, and even had time to polish what little silverware they owned. Somehow they lived to old age, raised a large family, all of whom went to college, survived prohibition, the Depression, two world wars, rationing, gender inequality, polio, TB, and measles. In short, they were like millions of Americans.

My grandfather never sought permission to live less hurriedly, less like a consumer, less oblivious to time. He never sought recognition for doing so. I would sorely like to have a little of his independence. I often find joy in walking the dog, washing the car, shopping for fresh vegetables, cooking dinner, or hanging up my clothes—the details of life. But somehow looming over me is a great, importunate world where existence and one's daily routine are measured out in nanoseconds. Sometimes I feel like just another petty functionary enmeshed in the classic carrot-and-stick game, trussed in my work harness, blinders in place, ready to go, eating from a bag tied to my head, defecating on the street while I work, running in place on my NordicTrack, apparently willing to work myself to death.

Educated as an engineer in Basel, my grandfather, like his father, became a man of the cloth, a Protestant preacher who immigrated to mid-America around 1910. His upright physical stature was a lifelong product of his early military training. Living to eighty-six, he walked two miles a day and was the most carefully groomed man I've ever known. Short, but with the body of a welterweight prizefighter, he was always dressed in a dark blue pinstriped suit and vest, starched white shirt, and pin-dot black tie set above patent-leatherlike, shined black shoes. He always smelled clean and of bay rum, with every hair in place and his nails perfectly trimmed; his breath smelled of the mint candies he kept in his trouser pocket.

Hardly a good consumer even in the prewar years, he made things last and last and last. It was typical of him to fold shopping bags neatly and use them over and over. He lavished lots of wax and repairs on his mail-ordered Sears shoes, which he wore for ten years. He kept his cars twenty years. He practiced fifty years ago what the grand old man of German design, Dieter Rams, preached only two years ago: "Less is better."

On a gold chain in his vest pocket, close to his heart, he carried his prized Swiss pocket watch, which his father had given him in 1896. I inherited the watch, and it still works perfectly. If my grand-

father were alive today, he would be aghast at the pressures to buy that are so much a part of our intensely acquisitive lives. To transform Sunday from God's day to Macy's day would have been a sacrilege to him, not so much from Christian rigor as from his notion that at least one day a week should be "a quiet one" (a civil prejudice still practiced in Switzerland today). The critical consciousness to "make good or green" choices in the marketplace and the confusion caused from sorting out value amid the welter of meaningless commodity choices would have been unworthy of his time or energy.

Later, after World War II when he moved from St. Louis to rural Missouri, he would rather have combed his chestnut horse, read a book, taken a walk in the woods, or written a thoughtful letter—by hand in ink with his inexpensive Esterbrook fountain pen—than visit a shopping mall. He owned hardly anything, save his oak desk, a rosewood barometer, an old set of drafting tools, a hoard of books, and a violin which he played rather well. Somehow he always had a shiny dime for me, and I thought him to be somewhat aristocratic.

In my lifetime, I will use up nearly 20,000 disposable razors, 25,000 gallons of hot water and 500 six-ounce cans of aerosol shave cream (sixty-two shaving years, God willing, one shave per day). By comparison, in all his years of removing whiskers, my grandfather used less than a half dozen straight razors, maybe two or three honing straps, at best a hundred bars of Ivory soap, maybe a dozen shaving brushes, one ceramic lather mug, and less than a quart per shave of hot water (seventy-five percent less hot water than I use). I can't help wondering if in such simple routines I shouldn't shun an obvious material wastefulness and indifference to the details of daily life and adopt a less material existence in general, extracting more pleasure from the art of living.

I know new houses built in Denmark can last—without major maintenance like a new roof—for more than ninety years (our common roofs in the United States last but fifteen years). The Danish over a hundred years ago reached a consensus among their breweries to adopt one common and easily recycled green glass beer bottle, and by so doing continue to save an enormous amount of needless product differentiation and wasteful packaging. I can hardly imagine Anheuser-Busch and Miller coming to such an agreement.

I know that commercial-grade office furniture and work tools outlast their residential counterparts three to one, that cross-country trucks have a useful life in excess of 500,000 miles between major overhauls, compared to the family car which hardly lasts 60,000 miles without expensive warranties or costly repairs.

Can the religion of universal consumption, the belief that acquisitiveness ensures progress, even partially be displaced by another dogma. What deeper, more meaningful social and environmental goals would sustain our living with less for longer periods of time and enjoying the process? To repeat Carl Jung, "The only sensible goal in life can be the increase of understanding and wisdom. Everything else is bunk."

I think that if we understood design, we could temper our acquisitive urges and concentrate on increasing our understanding instead of our insurance policy riders. I think we would begin to spend our money and our talents for the long term—and this is beginning to happen in rehabbing buildings and neighborhoods, belief systems, and automobiles. Replace obsolete consumption design criteria like extravagant speed, fuel usage, planned obsolescence, excessive energy costs, down time for expensive repairs, complex comfort features, and rapid loss of investment. Replace all that with a playful car a person can love, enjoy driving, adapt to new or old uses, update with the latest technologies, and maintain inexpensively. Create a car that increases in value with age, a car you and I would be happy to use for a lifetime, like a Steinway piano or my grandfather's pocket watch. Perhaps a car to pass on to our children.

Some cars in history have come very close to such life-conscious criteria, namely the Volkswagen Beetle (recently redesigned and reissued) of which 20,000,000 were made and the French 2CV, both of which were produced at reasonable cost over a period of twenty-five years without substantial changes in the millions of editions. More important, they remain two of the most loved automobiles in history.

Like love or a great friendship, our relationships with things that refuse to die for mere fashion's sake take on more meaning for us than our affairs with things that become premature junk. For a new toy to break forever before Christmas Day is over sends a message to the child who plays with it. Planned or arbitrary obsolescence brings with it a parting psychology. Maybe our hunger to remain forever young could be satisfied if everything around us wasn't vanishing into the atmosphere. It's natural to care for things, places, people, and good experiences. Sometimes when I go to New York, I ignore the latest cuisine offered in SoHo and eat at Billy's on First

Avenue where the old wood paneling smells like martinis, aged meat, and fresh seafood, and the waiter's name remains secret. This isn't nostalgia. It's simply a good experience that outlives change.

Ironically, at least for me, environmentalism is not the only motivation behind the idea of making things that last. It's civility or humanism. Time wasted, material wasted is not the only point of this story. It's more about living versus consuming a life.

7

Finding Civilization in the Wilderness

Thou shalt not answer questionnaires
Or quizzes upon World-Affairs,
Nor with compliance
Take any test. Thou shalt not sit
With statisticians nor commit
A social science.

—W. H. AUDEN,
A Reactionary Tract for the Times, May 1947

With great enthusiasm I took in a Marshall McLuhan lecture at UCLA in the early 1970s and the usual coffee chat session that followed. Here was a thinker and visionary to be sure, or so I thought. His bright description of a global village excited me; maybe we could bulldoze the walls that for centuries had separated cultures. Well, we have certainly knocked down a few walls since then—in addition to *the* Wall—and we watch the world with an intensity heretofore unimaginable. I can't quite bring myself to call our world a village yet, and I can't quite forget Robert Frost's reactionary farmer and his line about good fences making good neighbors.

The world is now "in our faces" all the time, no matter where we are. CNN has outdone the sun. It shines all day and night on Adak, Peebles, New Delhi, Thunder Bay, the rain forests of Brazil, and the streets of Los Angeles and Mogadishu.

Twenty years after McLuhan's lecture, I was in Kyoto, ignoring the television in my serene little room in an old *ryokan,* knowing that the brusque iconography of CNN would ruin my Japanese mood. A bit homesick after two weeks, hungry for a bagel and coffee instead of the usual miso soup for breakfast, I switched on the television for a taste of home. I got instead the first news of the Gulf War wrapped like a candy bar in CNN's slick graphic imagery. All at once I was in three places—Kyoto, Baghdad, and Atlanta. Everywhere and nowhere, thanks to the wondrous technology that sees all and tells all. I felt my nose being rubbed in the collective shame of Bhopal, Chernobyl, or Tiananmen Square. I saw Patriot missiles leaping into the skies over Saudi Arabia. The world had become a well-lit street corner under the watchful eye of television cameras. Is this what it means to be modern? Is this civilization? Is it for this that we have labored to invent and improve technology?

When I was a kid in the 1940s and early 1950s, there was only radio. I remember being scared at Walter Winchell's fulsome prediction of an inevitable nuclear attack from Moscow. Real terror remained abstract. World reality was only a black-and-white image on cinema screens from Pathé News, mixed in between Bugs Bunny cartoons and Humphrey Bogart movies. My older cousin had real wounds from World War II, but he wouldn't talk about them, at least not to a curious ten-year-old.

The really bad news in 1945 came either in the mail or over the telephone. In fact, if the phone rang after nine P.M., my family reacted with real alarm. If the call was long distance, it must concern something serious. Only bad news justified the expense of phoning beyond one's hometown exchange. There has been, and always will be, bad news. Bad news always affected us—personally, *really*. There seemed to be an impact implicit in the pace and intensity of its flow. Bad news traveled fast, but it wasn't "live." Can we coexist with "live" messages? Nurses and doctors who work in burn units of major hospitals can only stand a year or two of watching such excruciating human pain and misery before they have to quit or rotate to other duties.

We have assuredly become global—at least in the view we have from our living rooms, but perhaps not more civilized. Perhaps civilization can best be appreciated by its absence. It's ironic that in the midst of having it all, most of us find ourselves muttering, *I gotta get away from it all.* Sometimes my circuits shut down, and the only way I can reset them is to shut down the whole system for a while. I retreat north, to the woods, in much the same way E. B. White retreated to a small shack without electricity to write his wonderful essays. Of all places, out in the boondocks I've found a voice of civilization that operates on my psyche in much the same manner as a sunset over a still lake, a voice that clears away the rubbish from modern life.

Less than three people per square mile live here on the Minnesota-Ontario border. From this geographical perspective, I have more than once imagined a trek directly north along the ninety-two-degree longitude for thousands of miles. The likelihood of bumping into another human being is quite remote—all the way up through Ontario, across Hudson Bay, the Northwest Territories, across the North Pole, and south down the Central Siberian Plateau to Krasnoyarsk in Russia, the first city of significant size. Such a jumping-off place exists only here, or maybe in Tibet, or in the outback of Australia, or on the edge of the Sahara.

I'm writing this essay at the end of the road, literally. It's a seven-mile canoe paddle from the only road across large open lakes and through narrow gorges. My clapboard cabin, aptly named "the Revenge," nestles in a pine and white birch forest, on a mossy green carpet, atop a granite ledge with an intoxicating lake view stretching to a Canadian horizon. The cabin was a U.S. Army surveyors' shack at the turn of the century. By design it remains without modern amenities. No electricity, phones, indoor toilets, or modern appliances.

Nature is in charge here; this is a place of essences. The soft air smells of pine, the hazeless sky renders the lakes and forests a spectral brilliance. Birch and aspen leaves glitter like silver coins. Black-bottomed lakes reflect mirror images of the shoreline and sky. I can clearly observe the moon lying a few feet under clear water in the shallows. The friendliness of grouse and deer recall an earlier, collective experience of the world. White-throated sparrows sing "Sweet Sweet, Canada, Canada, Canada," while the loons sound like loons—crazy. The brief nights are blue-black, except for the starlight and the northern lights, both of which blaze in the sky. Only a few artificial decibels can be heard here—the echo of a canoe paddle against a gunwale or the throaty drone of old Beaver seaplanes overhead transporting adventurers, mail, and supplies farther north.

But for a handful of locals, great-great-grandsons and daughters of early Scandinavian loggers and Ojibwa Indians, few folks live up here in the crystalline winters (forty-five degrees below zero). Summers last but ten weeks. Falling out of a canoe in early May or late September, you risk death in water just barely thirty-three degrees. Imitation fur toilet seat covers are popular in outhouses to mute the shock of an ice-cold seat on a June morning.

Remoteness is real here, too, and it dramatizes potential dangers. It's hard to describe this feeling of danger in the midst of such beauty. It's not like being afraid of black bears or tornadoes or any direct threat. Rather, it's more like a shadow that follows you, the reality that to call for help is all but useless. Spots where someone drowned or got lost in a blizzard are remembered for years. No need to manufacture artificial thrills like bungee jumping here. Yet I love this place, I feel totally alive and well here. I dream and plan all winter long to get back up here after the ice melts in the spring, and I hate to leave in the fall. The annual cabin closing is a mournful task in early October. Although I rarely have a chance to spend more than a few days at a time here, this place has become a part of my being, it is with me wherever I go. This place sorts out the essences from the attributes of life. Ironically, it seems as if civilization suddenly exists where we humans aren't, a terrible state of affairs and one I hope we can change.

But this little essay is not so much about communing with nature; it's more about seeing public life more clearly by distancing myself from it and realizing that living in the city without access to nature can be unbearable. One tempers the other. Imagine Manhattan without Central Park. (I wonder if a Central Park would be created today in Manhattan. Probably not.) Or Chicago without Lake Michigan, or Venice without the Bacino di San Marco, or a room without a view. I love cities and public life, too. Strolling among throngs of shoppers down Michigan Avenue at night under a fresh snowfall, just before Christmas, is pure magic, as is listening to ancient instruments accompanying Spaniards dancing the Sardonna in a Barcelona plaza. The city, like the wilderness, reveals the essences of life.

But then a technological intruder reaffirms the bonds between nature and humankind for me. CBC (Canadian Broadcasting Corporation), senior to America's NPR, reminds me that civility and civilization consist of both. CBC is the only tether I have to the world of news and cars, technology and politics. As repugnant as the ubiquitous "boom box" can be on a street or in a city park, when it becomes the only connection to civilization, it takes on a special significance, as do other re-

minders that I live in a techno-world—a lightweight, carbon-fiber canoe, a flashlight, insect repellent, a pocket compass, and a book of matches.

On my battery-powered radio, from Thunder Bay and originating in Toronto, CBC comes in loud and clear. It's a welcome companion. CBC programmers know about the lonely vastness of the North. Many towns across Canada are hundreds, not tens, of miles apart. To be Canadian is to know space and distance even in the intimate cyberworld of today. Strange that it's the radio and not the television or the Internet that offers a hint of the real global village. Stranger still is the fact that the message is not in the media, but in the content.

CBC radio is so good perhaps because it innocently pursues Carl Jung's notion of progress: that only through the pursuit of wisdom and understanding, and not merely the betterment of conditions, can we arrive at a civilized view of life. Somehow, like "Sesame Street," CBC's daily fare is thoroughly engaging. It starts with the upbeat "Morningside Show," hosted by the peripatetic Peter Gzowski (now regretfully retired). His is a civil tongue and one that satisfies my hunger for civilized public discourse (a phrase that has since popped up repeatedly in the 1996 presidential debates), a scarce commodity below the Canadian border. As a listener, I feel honored by the respect CBC has for me. It's Tennessee Williams's "kindness of strangers."

Along about midday, other vibrant personalities serve up a rich smorgasbord of news, music, literature, humor, weather reports, and world and local politics. One of the CBC radio hosts interviewed an expert on flatulence by telephone from the UK, a professorial sort whose academic curiosity has been limited to the human body's inability to store gas indefinitely. He is the sole inventor of a fart meter—who says we have too much technology? This frivolous bit was followed with recitations from Shakespeare by Hamilton, Ontario, second-graders. At twilight I listen to the urbane, hushed musings of Lister Sinclair about classical music and poetry, a reading from Kipling intertwined with Handel. Finally, close to midnight, it ends, appropriately, with bedtime stories on "Between the Covers." It's so nice to be read to sleep. What is broadcast in between follows a flow of events just like everyday life, obviously planned, but rich in serendipity.

CBC is the only game on my AM or FM dial. This sense of restricted choice is a source of contentment, in my view, for it contrasts so strongly with the unlimited and unwarranted range of choices Americans face every day. After all, we don't really need ten different kinds of instant coffee. There is no sense of wasting my time, even when I have it to waste.

In the global village, there is much worth sharing and observing. Through slices of life like CBC's we can define our commonality and differences with civility and grace. Somehow CBC slows the sky's rate of falling; at the end of each day I get a sense of confidence that life will go on listening to CBC. It's true I want and need to witness the painful news of Oklahoma's wastings, or China's drought, or a local rape, but I also need to hear recipes for Rice Krispies squares and listen to Billie Holiday's blues.

The quality of our multimedia is the measure of our cultural deliverance. If all the North American producers of programming took it upon themselves to relate less to products of vogue, of hype, of voyeurism, of self-pity and anger, and rely on their own creative powers, perhaps we all could find the mirror of civilization we're searching for.

The quality of our public discourse helps shape our humanity, as does the expression of balance. Life does go on. The sky will be there tomorrow. And to see the skies of the Canadian north woods and those of Minneapolis as part of the same world is to step toward a civilized world.

PART II

Places of Civility

8

Art, Design, and Harlem Prep

The life which men praise and regard as successful is but one kind. Why should we exaggerate any one kind of life at the expense of the others?

—HENRY DAVID THOREAU, Walden

Art is the frustration of integration.

—JAMES MARSTON BATES

Germany, a little over sixty years ago, put aside artistic freedom for the national socialism of the Third Reich. We should not forget that artists like Max Beckmann, photographers like Alfred Eisenstadt, and designers like Walter Gropius fled Germany as a consequence. Art has never been openly suppressed in the United States where, after all, the irreverent spirit of pop art flourished. At the same time, art is not exactly embraced as an integrated part of our everyday lives. Artists like Christo are household names across Europe. One summer in Berlin, the Reichstag was temporarily wrapped in silver cloth—a product of his creative imagination. Ask anyone here about Christo, and, alas, you'll probably be told that it's something to fry chicken in.

While art and design are not exactly the same, art historians say that the essence of Matisse's paintings lies in their design qualities. Designer Charles Eames's chairs were regarded as works of art on their introduction in the 1950s and have become more so in the ensuing forty years.

We have only two major public art museums partially devoted to design in the U.S., and both are in Manhattan: the Museum of Modern Art and the Cooper-Hewitt. Milan is a veritable city of design. London and Stuttgart do not have just museums for design, but learning centers where children and businesspeople can find books and case studies on good design. Denmark's design culture is its primary export to the world. Switzerland's creative infrastructure is a model of environmental design, and in Basel there is the Vitra Design Museum, designed by the American architect Frank Gehry. Philippe Starck, for better or worse, is France's Michael Jordan of design. Taiwan is sponsoring research and design as a national priority. Unlike the French, who devote one percent of their tax income to the arts, Americans' small trickle of public support (less than one-half the price of one B-2 bomber) for the National Endowment for the Arts and Humanities has not only been reduced, but Congress has put out a contract on the organization's life.

Sadder than these facts, and what won't be reported in the news, is the fallout from these cultural wars: the further erosion of art curriculums in the schools (classes in music and art are the first to go in school budget reductions), design and research budgets are the first to be cut in business. Creating and renewing much needed parklands, before now a widely supported activity, has been squashed. Maybe we should be grateful we have as many parks as we do. Yet art critic Robert Hughes was able to write in the August 7, 1995, *Time* cover essay (a cogent defense of art's role in a civilized society, after a recent congressional attack on government's arts sponsorship), "Painters, dancers, actors [I would add designers, architects, and inventors] are tough as weeds and can grow in cracks in the concrete."

Art is a wonderful way to open our eyes, to show us confidence and *vérité civilité,* even while it depicts our pain. Art and design give us the skills to survive—both physically and spiritually. But many of us see art and design as luxuries, as leisure activities, things we would pursue if we only had the time. Design has somehow come to mean high-end goods only, "designer products." This divide-and-conquer–driven (by time and greed) business strategy not only undervalues the position of art in industry, it weakens the democratic traditions of design for everyone pioneered by Henry Ford, Thomas Edison, Steven Jobs, and Walt Disney. All but gone is any motivation to design and produce something inexpensive, durable, useful, and beautiful, like a Sunbeam toaster, a Sony Walkman, or a (full-size) Ford sedan circa 1950.

I do feel confident that art will out itself in America even in the face of conservative cultural warfare. America's only indigenous art form, jazz, grew out of human misery few of us can imagine—slavery, racism, and poverty. I'm convinced art and design will similarly triumph. In all of history there was and remains a wiliness about artists, scientists, inventors, and explorers who defied indifference and suppression.

I have been fortunate to see the power of art and design first hand. In the late 1960s, I helped transform a supermarket into a street academy. At Harlem Prep, a high school on 125th Street in Harlem, I learned that design attention can infuse spirit into people, activities, and places that have been "design ignored" too long. By living in the midst of a reconstructed supermarket cum private and publicly supported experimental school, I had to rethink any preconceived design concepts based merely upon traditional architectural concerns like style. Problems at Harlem Prep were defined by the failed rigidity of big-city ghetto schools, fragmented families, poverty, and unstable, depressed lives. Many students were in their twenties and thirties and had been previously awarded "jive-ass" (meaningless) public school diplomas. Many couldn't read, write, or do simple arithmetic; some students were without homes (a student who went on to Harvard slept on the subway every night for a week before graduating); some had drug problems; some nursed their babies in class. In all, 300 students had to be crammed into 10,000 square feet.

This was not a school where dress codes, dainty manners, chewing gum restrictions, or other behavioral niceties were celebrated. It was a noisy place; yet there was a palpable civility born out of an exercise in redeeming the human spirit the likes of which I'll never forget. The transformation of the building and the transformation of this piece of society proceeded hand in hand. Audible from the basement, on the street-level classroom floor, the heartbeat-like, rhythmic sounds of the "MoJo Logo" singers and dancers enlivened the serious subjects being taught. Our designer/student/teacher team hung student art and color photography lavishly around the school. We developed simple, multiple-use, user-changeable open classrooms that solved a host of untraditional school problems. Harlem Prep was a design challenge, but it reversed my usual comfort-motivated mission. It brought a measure of civil comfort to me. Design attention took on new meaning and commingled with the energy and spirit of suppressed but hopeful people. Like music, it comforted our souls. In fact, art and design turned the students of Harlem Prep into artists themselves. Their art not only hung on the walls, it allowed them to survive. They began to design their lives and their world, not simply experience it. There is art and design in a carpenter's or a waitress's work, in a scientist's theory, in a surgeon's operation, in a business executive's vision. We simply have to notice.

"Like politics, design is a local problem."—Ralph Caplan

9

The Red Door Clinic

You've been a good old wagon, honey, but you done broke down.

—BESSIE SMITH

Oh no. The secret scandals that gnaw at us and eat away at our self-respect are far worse. Are we Buddenbrooks the kind of people who want to be "tip-top" on the outside, as they say here, while choking down our humiliation within our four walls?

—THOMAS MANN, Buddenbrooks

The business of making us all witnesses has become a well-organized industry and a profitable form of entertainment. The media have turned us all into servants peeking and listening through a keyhole as the world unfolds in naughty, titillating ways. We ourselves are partly to blame for the fragility of our privacy. After all, we consume the stew served up in tabloids and on television. I sneak a look at the latest celebrity unveilings as I wait at the grocery store checkout. Secrets—deserving or not—are dished out daily to us, even as we truly mourn the deadly consequences of the apparatus designed to reveal them. Privacy, that piece of civility, is fast disappearing.

Nothing is simple. Sometimes I celebrate the absence of privacy. No doubt Rodney King's assailants were horrified to learn that their private violence had been permanently recorded on a video camera and made public. This kind of witnessing led to a measure of justice for King. The task force that nearly beat King to death for a traffic infringement exhibited their expertise via satellite to the world. The videotape spurred the indictment and eventual imprisonment of the police turned perpetrators. It is good, one feels, that such dark deeds are brought into the open. Perhaps fresh air and sunshine will help such cultural wounds to heal. Besides, we think, here is a citizen—the witness who made us all witnesses—who discovered the offense firsthand and stood up for decency and fairness. Maybe this person has struck a blow at civil apathy and uncivilization. But a lot of what is passing for news seems nothing more than voyeurism. Perhaps infotainment is congenitally unable to reveal the truth about reality. Our own experience teaches us more about human relationships than all of Geraldo's interviews with dysfunctional families.

We seem to be so fed up and angry over the moral frailty of our leaders that a goodly share of us can't wait to witness their public embarrassment. Yet, I really did not enjoy seeing Clarence Thomas and Anita Hill sweating nervously on national television. While it is critically important to know the moral underpinnings of our Supreme Court justices, one has to wonder if it is fundamentally civilized to employ the technology of public scrutiny so wantonly on individuals who will forever bear the aftershock of their public nakedness.

Although it has since been replaced by a parking lot and is now housed in more glorious surroundings, a public health service in Hennepin County in Minnesota in the early 1980s established itself without a message or sign that signified its location or purpose. Anonymity was part of its charter. This facility in the pre-AIDS era was hidden in a warehouselike building on the southern fringe of downtown, on a not-too-busy street and behind a red steel door. The clinic offered tests for sexually transmitted diseases at little or no cost. Few questions were asked, condoms and compassion were offered, in stark contrast to the hyped-up, pseudo-compassion of the talk shows.

Much of the population in Minneapolis has predominantly Scandinavian roots. Just about everybody has heard Garrison Keillor describe the kind of people who cannot reveal their innermost feelings, much less divulge the possibility that they may have a venereal disease or may have passed it on to someone else. I doubt this feeling is limited to a group of people well known for their reserve. Even today when sexual matters appear in all forms of common public discourse, I'm sure many people seeking a venereal or AIDS test, an abortion, or needing substance abuse treatment would be shy about going to their local doctors. Main Street largely remains Main Street, whether in New York, Minneapolis, or Bakersfield, complete with the fears and isolation brought on by sensitive personal problems. Certainly the rich and famous can afford to buy privacy, but even they flock to the anonymity of Minnesota clinics both inexpensive and expensive.

The all-but-invisible Red Door Clinic and its intensely private health care may seem humorous and a bit old-fashioned these days. Nevertheless, it was and remains an act of civility to provide discreetness

and privacy to everyone when it comes to our most intimate problems. Minnesota culture prides itself on its compassion for social failures like drug and alcohol abuse. Minneapolis, a short distance from the Hazelden substance abuse clinic, has a considerable subculture of recovering addicts. Many of these addicts find a host of aftercare services, jobs, and a forgiving atmosphere in which to seek more healthy lives.

I guess the point is this: We need to have compassion for human failure. Part of a real civilization, it seems to me, is a social understanding of human problems and a granting of the privacy that often makes them bearable. This understanding and compassion must be expressed in buildings and environments, as well as in attitudes and programs. That's where design and designers come in. That's where everyone can play a part.

The ability to find personal anonymity for whatever reason is a source of public civility of the highest order.

10

Rest Area and the Polysemous Side of Life

Just when I get the refrigerator paid for, the goddamn thing breaks down.

—ARTHUR MILLER, Death of a Salesman

Little doubt exists about the utility of hope, exuberance, and creativity in today's world. This, in the face of mixed messages about design, the environment—an entire civilization. We are also assaulted by the mixed messages of corporate advertising—the permanence of desire and the impermanence of what we buy. What are we to think when a sign on the parking lot attendant's booth says PERSONAL SERVICE IS OUR GOAL. YOUR ATTENDANT TODAY IS #54? Just when we believe we've found a solution to starvation in Somalia, we see our troops attacked and our country hated. Just when we've purchased a fuel-efficient car, gridlock on the freeway stops us in our tracks.

Being conditioned from early age to doubt our innermost feelings of worth, some children feel left out without the proper tennis shoes or haircut. Young teenagers have killed total strangers in the pursuit of a Chicago Bulls jacket. The local weatherman on TV has recently dyed his graying brown hair to keep up with the Ken and Barbie news anchors. The most painful measure of the mixed message is that which denies us individual propriety as we age or if we become disabled. Especially difficult is the mixed message in many products—technology in particular. The price is high, the life of the product is short, the residual value is nil.

Venturing out across the vast, limitless plains of our western states with our kids on a trip a few years ago, we rarely drove more than 100 miles before one or all of us needed a restroom. Somewhere, if memory serves me, in South Dakota or maybe Oklahoma, we came upon a sign that said REST AREA, 1 MILE AHEAD.

My wife, Sharon, said, "Be sure to stop, I just saw a rest stop sign." Within that short mile, all bladders went from passive restraint to immediate alert. Our family's usual response at times like this could be described as vigilant anticipation verging on sheer panic. Muted cries came from the backseat. The sign at the exit ramp said REST STOP/NO FACILITIES, and the station wagon filled with the quiet panic aboard a sinking ship. Somehow, everyone managed without the help of facilities. My children have never forgotten this comical moment.

Yes, there are well-designed, clean public facilities along much of our highway system. Wisconsin comes to mind, with well-landscaped rest areas, some of which are solar-powered, safe, and pleasant enough for a nap or a picnic. But what about cities? Metropolitan rest stops are often as welcome as ones on the interstate. In St. Louis, when public restrooms ceased to exist, my grandmother stopped shopping downtown.

While cities like New York have experimented with high-tech, self-cleaning public toilets on the streetscape, much of our public space is either devoid of bathrooms or dotted with drab, unsafe, and uninviting Porta Potties. What public toilets exist—in subways, depots, airports, malls, and parks—are abominable, unclean, poorly monitored, and increasingly hard to find. A lovely exception is San

Francisco, where handsome, oval, self-cleaning public toilets sit in parks and on street corners.

In Washington, D.C., recently, along the grand landscape of the Mall around the Washington Monument, the Porta Potties stood like embarrassing monuments to the most basic of public needs. Really, can't we do better? With all the glorious buildings and monuments in our nation's capital, isn't it time to enlist the talents of our greatest architects and build a decent public toilet? Strangely, judging by the minimalist bathrooms in corporate facilities and school buildings, architects seem indifferent to the simple fact that people need bathrooms. Except for a few in fancy hotels, I can't associate a well-designed bathroom with a designer I know.

In some countries it can be worse, and in some it's a matter of pride and civility. In Switzerland, the dogs have better toilet accommodations than humans in Houston, Miami, or Los Angeles. In Copenhagen, Sharon grabbed me by the arm to visit a public toilet she'd just visited. She said, "You won't believe this!" Not only was it impeccably clean, but it featured sheer curtains, gleaming blue ceramic

tilework, ventilation from an opened window, foot-operated sink faucets, incandescent lighting above real glass mirrors (makes a person look nicer than fluorescent lighting), a clothes hanger cum umbrella rest on the wall, and a beachwood bench to rest on—all in the men's room!

At least one measure of practical feminism has taken root in my home state of Minnesota. Though perhaps not as civilized as Switzerland or Denmark, our legislators took a small leap forward in civilizing restrooms and passed the "2 for 1" potty law. Henceforth, it is my understanding, in all public buildings, especially sports facilities, there shall be two stalls in every women's room to every one in men's rooms. And all the men's rooms shall have diaper-changing tables. Of such small things is civility composed.

Prosaic as it would sound on a national design agenda—less glorious than designing cars and computers—merging modern technology and age-old notions of civilized existence in bathrooms is certainly a job for designers. American manufacturers regard the toilet as a fashion object only, with its archaic functionality untouched for decades. Its operation hasn't changed in this century, only its form and availability. I was part of a research and design endeavor (the Metaform Program, sponsored by Herman Miller, Inc.) in the late 1980s that through the study of older and disabled people resulted in a toilet (designed by Gian Zaccai) that was height adjustable (for children and the elderly), provided armrests for ease of entry and exit, featured the combination bidet-toilet-paperless hygiene system (new to America but common in Japan), and that was self-cleaning. Designed more for private than public use, this technology is not my fantasy. It's possible, and it's immediately applicable in public restrooms.

Investments of this kind in public versus private facilities will go a long way toward reestablishing civility in public areas. In the case of public needs, we could use more technology, not less. We can reconcile spontaneity and predictable human needs with hygienic and esthetic values.

There is no reason our society cannot come to grips with the both/and of civilized and functional environments that account for basic human needs. It's hard to appreciate the Rocky Mountains when you're desperately looking for a bathroom.

11

Water Works

Big ideas and major public messages are delivered in subtle and small ways through the quality of the environment. I think of the old horse troughs in every Swiss village and even in some cities like Basel and Schaffhausen. Horses don't often need a drink in the center of town these days, but the Swiss have carefully maintained hundreds of these equine drinking fountains, some of which date back to the 1500s.

In Basel, at the intersection of five major streets, next to a trolley stop and news kiosk, sits one of these drinking troughs, a modest architectural monument to civilization. Nothing grand, mind you, with geysers or spouting stallions, but a large bowl, immaculately clean, a foot or so deep, replenished by a small trickle of water dropping from a pipe extending like Dizzy Gillespie's trumpet from a chubby stone face.

The message here is as clear as the water: "You are part of a civilized society. Our water is pure and

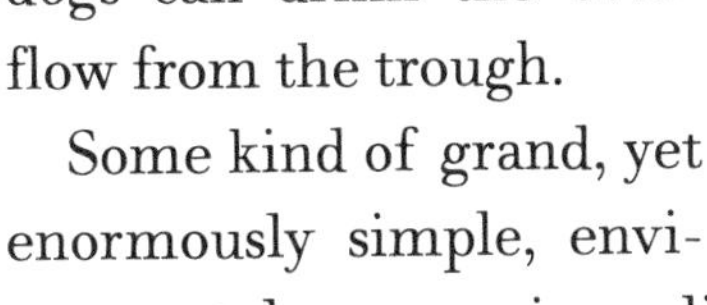

clean." It trickles down from the Alps and finds its way to all living things. Bicyclists splash it on their sweaty faces; pedestrians cup their hands for a drink. Most charming of all is the bowl at street level where dogs can drink the overflow from the trough.

Some kind of grand, yet enormously simple, environmental message is on display here. This is water works of the best kind, not a tumultuous or bizarre imitation of nature; not water as something to toss coins into; not water as white noise for some shopping mall—but simply an age-old manifestation of civil goodness, a simple and pure message about nature's most profound element and its relationship to life.

Free public access to water was one of America's basic amenities before our paranoia and pollution and the smell of chlorine drove us to bottled water from France, and Italy, and Canada. To see trucks hauling Perrier and Evian water to Aspen and to watch people on streets there consuming it is to witness a serious decline in the quality of our civilization and our lives.

While most of our tap water remains clean enough and safe enough for drinking, many of us worry that it is laced with poisons. Sometimes, as the people in Milwaukee and Washington, D.C., found out recently, it is. Public drinking fountains, once a common part of American parks and cities, are increasingly rare. The quality of life and the environment is measured in the smallest of gestures—a sip of good, clean, cold water or a place to sit down.

To preserve civility and even our civilization, we must become aware of, and attentive to, the great meaning in small gestures. We must make these gestures and guard carefully the resources that allow us to make them.

12

The Creed of Lake Harriet

Civility being nothing but a certain Modesty and courteous disposition which is to accompany us in all our actions. . . . Civility is a Science that teaches us to dispose our words and actions in their proper and just places.

—ANTOINE DE COURTIN, Rules of Civility

I have a respect for manners as such, they are a way of dealing with people you don't agree with or like.

—MARGARET MEAD

Americans are so accustomed to less-than-genuine greetings in public that to say "Have a nice day" or "Thank you for shopping at KMart" now holds virtually no meaning whatsoever. Even when rudely confronted by the red slash in a circle (which derives from European NO PARKING signs) forbidding us from doing this or that in otherwise friendly surroundings, Americans take it in stride. Or do we?

No one would argue the importance of a stop sign

or the necessity of a sign directing us to a public toilet. But the manner in which these messages are communicated is a matter of civility. The way we perceive them, the way they litter or grace our world, affects in some ineffable way our desire to be part of a civilization and not trains in a stockyard.

While a guest at a stately Scottish inn west of Edinburgh, Scotland, near the town of Peebles, I was struck by how much the way in which these little messages are delivered can affect behavior. Sitting down to breakfast in a somewhat formal dining room, I noticed a small folded white card immediately in front of an elegant place setting, complete with fresh flowers (not plastic substitutes), which read, "If you can refrain from smoking in the dining room, this would be appreciated by other guests." I smiled and was amused, and I pondered the message beyond the words. It sunk in.

On one hand, the message insulted me, implying that I might not have the will to withstand the temptation to smoke with my breakfast coffee. On the other hand, it touched me to think someone was sincerely trying to treat people with respect and understanding. I hope that the authorities had an even higher purpose in delivering their message in such a way. I think—and hope—that they were aware of how much pleasure some people find in lingering over a pot of coffee with a cigarette. I hope that they meant to say, "Gee, if it really is important to you, go ahead and indulge yourself, we'll manage quite nicely with the smoke, thank you very much."

Oh, come on, you might say. Why isn't a simple "no smoking" good enough? And what about the people who find pleasure in lingering over a pot of coffee *without* a cigarette? The point isn't whether you smoke or not. It is that somehow we have lost the sense of toleration and understanding—of accommodation. (Remember, Alfred North Whitehead shrewdly observed that there would be no such concept as toleration without behavior that some of us find intolerable.) Surely we have other alternatives to sorting smokers into ghettos in public places. Surely self-righteousness about smoking—or anything else, for that matter—does nothing for anyone's sense of civility or community.

To me it is a matter for design to make a community in the face of divisive attitudes. We need to design our attitudes and behavior. It would be too easy if I were able to design an object or a building that generated community. No designer I know is that smart. Yet community used to be designed very well, thank you, without designers. Community was created without urgings by the First Lady or the theories of sociologists. Here is one expression of bygone community spirit. The plaque bearing this creed is not a mile from my own house.

Creed for the Harriet District

I believe in the Lake Harriet District.

I shall speak a good word for the Lake Harriet District every time I get a chance. There may be some things about the Lake Harriet District I do not like, but I am not going to pass them along to everyone I meet; rather I shall enumerate the good features of our Community, and they are many.

I believe everyone should consider himself a Committee of One to further the interests of the Lake Harriet District, and I will try to be such a one.

I believe if I cannot say a good word for the Community, it is the better part for me to keep silent on the subject.

I believe there are many who could say a good word for the Lake Harriet District without adding to their expense account.

I resolve that I will do all I can to make this a Community of good report and creditable appearance. Believing this can be accomplished only by every individual doing his part.

I faithfully promise to assume my part of the responsibility.

Reading the Creed of Lake Harriet is to reflect hesitatingly upon the cobwebs of Victorian virtue and a very real sense of community. Seen in today's "tell it like we think it is—tell it all and tell it loud!" tabloid world, the very notion of not saying anything bad about a place if you can't say something good seems hopelessly naive, romantic, and nostalgic. Yet deep down inside this expression of the "Minnesota Nice" syndrome (I've heard that other states claim this phenomenon) lies an element of civility we are losing in our cynical, late-twentieth-century rush toward the next millennium.

I won't even try to explain why or how the "Minnesota Nice" phenomena exists, but it does. It's no myth. There exists here, where I live, in south Minneapolis near Lake Harriet, ten minutes from downtown, an inexplicable level of long-standing (not neo-communitarian) neighborhood tenacity, an admixture of the old and the young, single parents, traditional families, racially mixed families, gays and lesbians, and flocks of golden retrievers.

We have relatively low crime and violence levels, we intensely use our beloved Lake Harriet in all seasons, no matter how cold. We pick up dog droppings with dispatch, our small businesses prosper without franchises, we are involved in our good public schools, we carefully restore and maintain our century-old, eclectic domestic architecture (of middle-class affordability), we have the last mile of a once-great trolley system that reminds us of better urban forms of transportation than our grid-locked freeways, and we put up with the noise of jet planes overhead. We live and practice Minnesota Nice. And we often feel like we live in a civilized world. Now that I've purged myself of all this self-congratulation, the question is why does it work.

Maybe we inherited it, although I grew up in St. Louis, and a good share of the neighbors are from California, New Jersey, Ohio, and elsewhere, places where Minnesota Nice is heartily scorned. The Minnesota Nice stories are legion in these parts. Minnesotans are said to believe that "everywhere else, the grass is always browner." As a sign in the Twins' dome proclaims unabashedly, WE LIKE IT HERE!

Lake Harriett Band Mell

Minnesota Nice manifests itself in more than cozy, upper-middle-class neighborhoods. Some years ago the Minneapolis *Tribune* published an article criticizing the city's overly generous welfare policies and printed a story about two West Coast bums who ride the rails 2,000 miles every summer just to get free dental care and new eyeglasses in Minneapolis-St. Paul. World-famous treatment centers and half-way houses and shelters exist here. Cynics will, and do, say somewhat truthfully that Minnesota has no *real* problems because of its homogeneous, essentially white Northern European population, compared to the pot-boiling cities of Detroit or Chicago or Milwaukee. Or that the Twin Cities (though twins they're not) are small potatoes—small towns masquerading as big ones (ranking only eleventh in population nationwide). Or that having never been industrialized properly, this region has escaped postindustrial decline. But it seems to me that the signs of civility I see here result from more than an historical accident. And despite my enthusiasm, civility is not unmixed with heart-breaking despair.

If outside observers and home-grown parochials would dig deeper, they'd see the stubborn downward spiral of the urban American Indians, the increasing numbers of jobless and poor rural immigrants (many from the depleted northern iron-mining regions), or the major rise in rapes and murders, reaching record highs in 1995 (at a rate of increase higher than that of New York City), or the thousands of ghetto refugees from places like Peoria, Illinois, and Southeast Asia. Oh, yes, right here in River City they would see many of the same things they see elsewhere.

The problems we have will not be solved by major-league sports or the lottery. I happen to think that the civil society of our cities will survive with a dollop of soulful naughtiness, a laid-back toleration for human frailty, and a healthy measure of ownership. Too much Minnesota Nice is not good. It becomes the vice of our virtue. There is a fledgling *soulfulness* here which, in places like New Orleans or Chicago or L.A., seems to increase civility and tolerance. It acts as an antidote to too much sweetness.

In Ann Arbor, Michigan, a place much like South Minneapolis, Clark Malcolm and I took in a blues concert recently. Two African-American friends of his enjoyed the predominantly white musicians' blues but couldn't help saying that they were expecting something "more down and dirty." We knew what they meant. The music was too much Michigan Nice. Diversity, eccentricity, naughtiness, the oddball—civil societies have always contained these elements outside the mainstream. The southern blacks who came to Harlem found New York's cultural climate liberating (at least artistically). Those who immigrated to Detroit found the utilitarianism of its assembly lines just another form of servitude. Somehow I believe New York was more civilized than Detroit. Business came first in Detroit, and the soulfulness and weirdness that impeded business was not welcome. Many people here in Lake Wobegon-Harriet recognize the good things about searching for civility in cities. Maybe being adrift out here on the edge of the Great Plains gives us a great hunger for growing our own art and diversity. It's just too far away from traditional cultural oases like Boston, New York, and San Francisco, cultural centers first and places of business and enterprise second. Smug as it sounds, no one I know here has any longing for business-first places like Houston, Orlando, Detroit, or Branson. Commerce, after all, is only a means to an end.

Despite our problems and the threats to civility in living in the times we do, we may after all find and preserve a modicum of civility in Lake Harriet. Prince and Jimmy Jam live and work around here somewhere, and even Garrison Keillor has come home again. No matter how global we think we have become, we are tied to the here and now of our homes, apartments, cities, and neighborhoods. We are parts of the local scene. Or we should be. I think "global community" is somehow a contradiction in terms.

Have you ever been tempted to slam on the brakes in a rented car? To drive it a little more recklessly than you would drive your own? Why not, I sometimes secretly think. I won't have to pay the maintenance bills. Perhaps this is a trivial sin, but maybe it isn't. I tend to think that we treat our entire world like that—as if we had leased it for a few years and will eventually exchange it for a new one. We all know it doesn't work that way.

Sometimes I think we've become a nation of renters, here for the short term, whether *here* is a place of residence, a stock portfolio, or a political position. We have lost a sense of proprietorship and the sense of community that tags along behind ownership. If we aren't the proprietors of our civilization, who is? (I must add—for the sake of my friends in large urban areas—that I don't consider people in New York who rent apartments for decades *renters.*

I know that the chairs I design will have owners, even if the owners work for organizations that bought them the chairs. Design functions with the premise of use. It's impossible for me to conceive of designing anything in the abstract. I have been able to make a living as a designer because people own things.

What is more bleak than a vacant lot? Part of the standard description of the urban nightmare includes "vacant lots filled with rusting cars and rotting garbage." They are vacant and depressing precisely because nobody owns them and I don't mean *own* in a legal sense. I mean it figuratively and spiritually. Yet when communities have decided to own vacant lots, they have turned them into parks, playgrounds, gardens, all of which speak to civilization and work against dissolution.

I often wonder why we act like we rent so much of life and reality. Have the world and our lives become so complex and unfathomable that we shy away from owning anything lest we burden ourselves with too many problems and too many obligations? Don't we revel in Washington's victory at Trenton on December 26, 1776, partly because he beat those rented troops, the Hessians?

Renting removes us from the responsibilities—and the joys and benefits—of ownership. The renting attitude is seeping into our relationships with cities, neighborhoods, and even people. We rent a friend exactly as we rent a car. We even rent a world every time we use a nonrecyclable plastic bottle or open another strip mine or send another species into extinction. We are the owners of our world, and the sooner we step up to our responsibilities in that regard, the better.

Civilized communities are places to be longed for and to come home to. Civilized communities, like well-designed products, mediate extremes without destroying them, encompassing worlds in small spaces. In civilized communities, every person assumes part of the responsibility for life there.

13

Beer and Bacon Sandwiches at 5:30 A.M.

And may her bridegroom bring her to a house
Where all's accustomed, ceremonious;
For arrogance and hatred are the wares
Peddled in the thoroughfares.
How but in custom and ceremony
Are innocence and beauty born?
Ceremony's a name for the rich horn,
And custom for the spreading laurel tree.
—W. B. YEATS, "A Prayer for My Daughter"

No occupation is so delightful to me as the culture of the earth and no culture comparable to that of the garden.

—THOMAS JEFFERSON

Many believe the twenty-first century will bring via technology a general substitute for *real* life. "We are standing on a revolutionary threshold," says MCA's Edward Teller, a mastermind of the forthcoming cyberworld. Danny Hillis of MIT goes even further, predicting a kind of technological Darwinism that will self-select the kinds of artificial intelligence and technology to survive. Artificial intelligence will become self-perpetuating, ungoverned by human moral choice. Future existence may be a life, if we can call it that, devoid of material need wherein all experience will be artificial. Woody Allen's orgasmatron come true.

From Icarus to Frankenstein to Johnny Mnemonic, humankind persists in seeking ways to redesign itself. There's much speculation as to why. Many people speculate as to the reasons; many peo-ple never even wonder why. Perhaps our hunger for immortality won't be satisfied until we can outdo God. It's an intriguing possibility to think that at some stage in life, we can *redesign* our basic features: download more memory and IQ, trash out bad memories, install Pavarotti-quality techno-vocal chords, add or subtract stature or body proportion and shape—at my weight I need to grow five inches—chemically alter skin color, perk up our breasts, smooth out our wrinkles, backwash the tar from our lungs, de-shard our arthritic joints, postpone or totally eliminate our date with death.

So successful have been our recent attempts to outwit and supplant nature that machines and the human body and plants appear to be one with the other. Artificial intelligence and artificial experience now symbolize technological progress, just as automation did fifty years ago and horsepower a hundred years before that. All of this excites me as a designer, but it also poses a problem. Somewhere in the blurring of the boundaries between machine and flesh, between molecules and visible matter, the sought-after alchemy between technology and art rarely happens, even when the forms become identical. Without a footing in reality we wind up with *ersatz* life.

Confusion reigns today about what's real. Nanosecond technology beckons while my heart beats at the same tempo it did thousands of years ago. Worse is the general discounting of states of community and nature in favor of extracts thereof. What was good in the past becomes a nostalgic extract in the present: small-town life memorialized in chic shopping mall architecture mimicking historical styles; Disney-juggernaut visions of the future city with monorail transportation and immaculate streetscapes and without trash and the smells of life. Living with the negative fallout and unfulfilled promises of past utopian visions, I remain cautious of cyberworld utopias. I'm amazed at the good things that no one predicted, like the laptop computer or the demise of communism. Wouldn't it be more useful to shape up *this* planet for the next tenants? Design deals with future possibilities, sure, but welded to present reality.

In the crucial pursuit of extending the shelf life of tomatoes or sweet corn, why is chemical technology enlisted as the most immediate and desirable solution, as opposed to encouraging local agriculture, food distribution systems, and food markets to deliver the real stuff in a fresh, unadulterated state, on time? Those of us in the corn belt where fresh corn should be seasonally available (we can still appreciate the joys of a juicy ear of corn) not only dislike the intense sweetness that injected fructose (sugar prolongs freshness) brings to this food, we resent the unabashed acknowledgment that our food is being tampered with—and for purposes other than flavor, texture, or wholesomeness.

It is not objectionable for science to *try* to deliver an ear of sweet corn rich in essences? No. It just seems ludicrous that we try so hard at the expense of other measures, and that the results have become part of what in my view is a growing vocabulary of needlessly extracted real-world experiences, be it sweet corn or sex or travel or communication.

Worse, under the guise of convenience, extracts of experiences and materials are offered up as fodder for the middle class and middle enlightened. Extracts are a source of huge profits. Food technologists can make a twelve-ounce can of soft drink from a drop of real fruit juice. In terms of natural food for the table, or natural fibers in clothing, or natural materials in housing or furniture, it's the quality differences between the real and the artificial we accept too easily. Around here in Minneapolis, these differences are humorously characterized as "the pork and beaners versus the L.L. Beaners." Pork and beaners shop at the warehouse marts and don't eat free-range chicken or naturally sweet corn. The L.L. Beaners shop at the farmers' markets and co-ops and eat radicchio.

I hope that the enlightened and rich won't be the only ones able to afford the "natural" in the future. We are not mere victims of the ersatz aspects of life but the perpetrators of them. We demand fresh flowers and ripe tomatoes out of season; we have willingly exchanged the attributes of convenience for the essences inherent in the four seasons; we depend on shopping twenty-four hours a day, seven days a week. Perhaps we have lost the patience to wait for the harvest.

Somehow, if biodiversity is a good thing, perhaps experiential diversity is worth preserving, too, particularly by design and by designers—who should be the guardians of good experiences. Is flying interesting anymore, or playing baseball in domed stadiums, or highway travel in auto-controlled vehicles, or recorded MTV a general substitute for live music, cyberporn a substitute for real emotion, or is being cocooned in hermetically sealed offices communicating with bits a worthwhile replacement for face-to-face encounters, or salesperson-free shopping a social experience? Admittedly, I am inclined to exaggerate the drawbacks of the artificial. Nature has its dark side, too. I can't see without artificial lightweight plastic eyeglasses. I can't live through the pollen season without antihistamines. Past sixty, I'm not willing to give up my polluting, noisy snowblower.

Being "online" is a way to reconnect our fragmented societies, to extend intelligence and understanding. But the cyberworld shouldn't be all there is to look forward to. William J. Mitchell in his recent book *City of Bits:* "When attached to a display device (like a television set or personal computer monitor), such an appliance presents itself as a hearth that radiates information instead of heat. [I've always enjoyed communicating with a real fire.] Just as the fireplace with its chimney and mantel was the focus of a traditional living room, and later became the pivot point for Frank Lloyd Wright's box-busting house plans, so the display—the source of data, news, and entertainment—now bids to become the most powerful organizer of domestic spaces and activities. In most rooms, it's what most eyeballs are most likely to lock onto most of the time." Soon we will be celebrating Christmas not around the hearth but around a tubeful of CD-ROM images of burning logs whilst sipping a low-fat, multichemical-laden eggnog.

Reality has an odd way of winning out over technology in the long run. Despite some hundred-odd artificial parts and assemblies for replacing original parts of the body, we remain flesh and blood and eventually die. All the prosthetics of modern life, the substitutes for reality, don't make us more human, though they may ease our lives or even allow us to live longer.

I'm no Luddite. Finding myself in love with nature and technology, I am increasingly lonely on an isthmus between opposing and turbulent seas, one technical and efficient, the other artful and pleasurable. Obviously, this loneliness didn't start with the cyberworld. Truth is, much of modern science is doing artful and wonderful things. People will never return to some primal, pristine, and noble natural state, if such a state ever existed. But unlike those who exaggerate the promise of the artificial and who seem to be at our future's helm, I am disposed to reexplore our relationship with fresh sweet corn, down pillows, open convertibles, real grass ballparks, trains, and the idea of making the environment a child's garden to play and work within.

I doubt that a computer program will ever make civilization in our world run on its own. I doubt that we will ever be able to download into our children a knowledge of the arts of living without living real lives with them. I want some way to suggest that we can live nonviolent, enduring, and civilized lives within the constraints of our society. Since I happen to be an industrial designer, I naturally think that design can help.

Consider the design of the agri-industrial distribution complex. America is well fed with the off-season food flown in from all corners of the globe. It's not unusual to find fresh strawberries in January in Detroit supermarkets, and there are also various

Old Soulard Market, South St. Louis

ethnic food shops, a few bakeries, and in some cities seasonal farmers' markets abound. Even small local breweries are in fashion, accounting for the only growth in beer sales while the mega brewers' sales remain flat.

What's wrong with me? What's lacking? A genuine urban food market, that's what—a market like the Soulard Market in south St. Louis. I was weaned on fresh fruit, home-grown poultry, and vegetables and I never heard about the likes of "free-range" chickens or mesclun lettuces. Like Covent Garden in London or Les Halles in Paris, the Soulard Market was the belly of St. Louis—not a Fortnum and Mason or Dean & Deluca, but a food emporium without pretense. It was a visceral connection between the regional garden or farm and the table that doesn't exist much today except in rare leftovers like Detroit's Eastern Market, where you never buy a raccoon without its feet (so that you can be sure of getting a raccoon and not a dog).

While Soulard barely survives and is increasingly marginalized by tourism, and Les Halles has been leveled, replaced by more modern and strategically located food distribution systems, I can't believe that the great presence and personality of these markets aren't worth reviving. Gone in most of America and in these European cities is another facet of civil life—a vibrant celebration of quantity, freshness, and variety housed in significant architecture; a design theater for food and its essential connection to everyday folks and the fecundity of nature—the smells, flavors, and textures of the sea, the rivers, lakes, and farms smack dab in the center of life. What we have are thrice-removed, efficient, clean, modularized, systematic, overly decorated, smell- and taste-free supermarkets.

Super, they aren't. True markets, they aren't. But convenient, they are. We have what passes for fresh strawberries year-round, Norwegian salmon, Hawaiian mahimahi, and esoteric fungi. We can also have test-tube tomatoes hard as rocks and tasting like some bland biomass of unknown origin. We have California rice completely devoid of flavor. We have bottled water from France. Often we have no idea where these foods were grown or at what expense to Third World producers. Nor do we seem to care. Strange to me, while I can select papaya from Hawaii or grapes from Latin America, I can't get whitefish or golden caviar harvested from Lake Superior, all of which bypasses our local markets on its way to Sweden. The same can be said for local apples, trout, or chicken. Something is funny here. And let us not forget that this "problem" doesn't even exist for many people in the world.

I'm spoiled when it comes to food. I'm fond of fresh produce shipped in from warmer climes when I'm deep in the Minnesota winter. What I object to is the near total disappearance of local food producers, the loss of merchant personality in the stores, the flattening of flavors in basic stuff like chicken and tomatoes, and the faddish elitism currently attached to plain and good-quality foodstuffs. Somehow I want to keep Martha Stewart out of my life and out of my kitchen, in spite of her civilizing influence on the arts of daily living in America.

It doesn't have to be like this. In 1976, I was treated to a tour of Covent Garden by Michael Green, a friend and English designer. It started in the gray dawn light, at 5:30 A.M., with a beer and bacon sandwich in a local pub frequented by food merchants. I'd never smelled the odoriferous joys of fried English bacon and yeasty beer together and I'll never forget it. Later, as I walked into the market's flower house, my senses soared. Soft light fell from the crystal palacelike cathedral ceiling. Birds chirped in stands of flowers of all colors, varieties, and perfumes. Workers with flowers in their lapels seemed as cheerful and bright and fresh as the mums and daisies. In the middle of London's bleakness, this place was like a new morning in a great urban greenhouse. The structure was connected to the fresh produce hall, which had a similar lacy architectural canopy—light and delicate. The smells of the farm, the earth, and the scents of numerous familiar fruits and vegetables hung in the moist air.

The pace at the Billingsgate fish market nearby was much faster, almost nervous. Stacks of wooden boxes with labels stenciled on their sides noting fishing villages like Portree on the Isle of Skye stood filled with iridescent fish fresh from the ocean, enmeshed in jewel-like crushed ice. As it slowly melted, the ice created a wet floorscape that glistened like rain on a city street. The place smelled of seaweed.

The meat market was yet another personality and sensory experience, appropriately quiet and somber. The soft sawdust I walked on muffled sounds. Great racks of ornamental steel, painted a deep green and suspended from an architectural bridge, contrasted with ivory white ceramic brick walls. The contrast enlivened the pink skins of the dressed pigs, and the deep red sides of beef gave a theatrical presence. The end state of the slaughterhouse was more than a slab of meat pressed into a Styrofoam package. Here it was a celebration of meat marketing.

The question is how can we make design theater work in the health and hygiene-obsessed United States. I don't think it would be difficult. My urban garden market would be designed to remind us of the sources of things. Yes, it would have to be convenient, but the single most important concept would be an old-world sort of integration—of socialization, learning, playfulness, and shopping. I like to be entertained when I shop, but I abhor artificial entertainment in the form of playgrounds and clown motifs. I thoroughly enjoy the natural variety of a great marketplace. Good shopping for good goods in a good place is its own reward.

My garden market would be a place no larger than a typical Kroger or Safeway but located in a neighborhood and not in a mall. Yes, even in suburbia, I'd tear down some houses and plunk it right in the middle of residential life, where kids and old folks can walk to the store. So located, it needn't have acres of blacktop parking, but a reasonable few parking spaces for those addicted to driving short distances. It would have the most modern bagging and checkout services, and public bathrooms and drinking fountains. Like churches or fire stations, both of which are welcomed in suburban neighborhoods, my market would be housed in a domestic architecture, so that neighbors would not object to it on esthetic grounds. Vast amounts of natural light would stream through the ceiling, reducing the need for the high-intensity lighting of today's windowless markets. At the center of its cruciform plan, a small tearoom or coffeehouse would feed shoppers homemade pastries, good bread, and those marvelous half sandwiches, good pickles, and real soda fountain treats. Maybe there would be a pot-bellied stove. Wide aisles along each axis of the building would be edged by small independently owned and operated shops. Direct-source poultry raisers, fish purveyors, sausage makers, spice and herb specialists, grain and rice sellers might permanently set up shop. Perhaps the marketplace could be community owned. It could not be franchised; nor would it enlist subfranchisers like Starbucks or Taco Bell. Small kiosks would sell fast food, good old American red hots, hot pretzels—the chewy, salty ones.

Small domestic birds and rabbits would roam around, so that kids could learn that animals live in places other than zoos. There would be some cages with singing canaries and finches. Maybe a functioning hen house, so kids could see where real eggs come from. No artificial music allowed, but local musicians would be welcome.

Above all, the urban market would be a place to meet one's neighbors and reengage in the bountiful fruits of the gardens of America.

14

Staying Put

The things of the world have the function of stabilizing life, and their objectivity lies in the fact that . . . men, their everchanging nature notwithstanding, can retrieve their sameness, that is, their identity, by being related to the same chair and the same table.

—CHRISTOPHER LASCH, The Minimal Self

A thirteen-year-old said, when asked by a reporter from the Minneapolis newspaper where he was from, "I don't know, I've moved thirty-three times." Why do Americans move around so much? Can a civil society take root in the midst of constant mobility and change? Can you be a New Yorker without New York?

America is for me Home Sweet Home, for all the usual reasons: artistic and religious freedom, the Bill of Rights, economic opportunity, diversity of race and ethnicity, lots of space and fewer people per square mile. I can even live *with* Las Vegas (my son and daughter-in-law live there, after all), Disneyland, Mount Rushmore, Branson, and McDonald's. How can you not like a place that invented baseball, the T-shirt, movies, the pursuit of happiness, and hamburgers.

A native New Yorker friend asked me, "As a designer, obsessed with form and technology, why do you live in the untrendy farm country of the Midwest, and not in the artistic milieu of New York or L.A.?" Paraphrasing Studs Terkel, I answered, "Trends may start on the coasts, but the real problems start in the Midwest." He told me he would take New York any day, because, he said, "I can get a banana malt at three o'clock in the morning."

He's right, of course. That's why I really do like New York, too. But I stay in the Midwest because I was born and raised here, in St. Louis, near the Mississippi River. I still live near the same river because I feel connected here, because for me a connection to the past is part of a civilized life, and a geographical rootedness gives a certain stability to my life. It allows me a certain perspective. Like most of you in search of an education or work, I've been duty-bound to go where the action was over the years. I've uprooted myself and my family on fifteen separate occasions pursuing work from coast to coast. Yet here I am, back in the Midwest and aiming to stay here from now on.

In America it seems the rich stay put and the rest of us move from place to place all the time. They may have houses in Manhattan and Aspen and Sanibel Island, but they rarely move their homes. They can afford the grace to nurture connections to communities, clubs, institutions, banks, and governments. It used to be the other way around, you know. The serfs stayed on the land, no matter who the owner was. Now, the wealthy travel and are highly mobile, but keep real estate real.

Of course there are good reasons why people of average or humble means find it necessary to move their homes, usually a job. But there is a price beyond that paid by uprooted schoolchildren and trailing spouses. If you think you may shortly become transient, you will avoid much eccentricity in the outward propriety of your homesteads—your house will have to look noncontroversial, be ready for Century 21 to buy or sell quickly, like system barracks in the military. And so all our suburban castles must look the same—whether in Seattle or Atlanta or Cleveland. This enforced sameness erodes our individuality and gives our houses all the uniqueness of peas in a pod.

It's little wonder that even those of us with jobs fail to develop strong communal relationships, given that we live under constant threat of eviction by downsizing. The stories of places like Pittsburgh and Akron are cautionary tales for us all. People there saw their communities devasted by the demise of big industry, steel, and rubber. Since then, the citizens of Pittsburgh, where I had my first design job, have done everything possible to change their city from a utilitarian steel town to one of cultural diversity. They've spent billions as insurance against special interests controlling their destiny. The same can be said about Akron, Seattle, St. Louis, Minneapolis, and Denver. No single industry can destroy the fabrics of life these cities have woven by going out of business or pulling up stakes. With much of the bicoastal cultural hegemony breaking up and dispersing into the rest of the world, smaller regional chunks of America are taking their cultural responsibilities to heart. They are creating the same conditions for civility, on a smaller scale, that used to reside only in a few places.

Culturally, America is trying to stay put and is consequently changing slowly and for the better. By staying put, populations are investing their communities not only with cultural meaning, but also with a modicum of economic stability. I can't get one of those banana malts in the wee hours in Minneapolis yet, but I can walk my dog at five o'clock in the morning. And I can certainly come to experience and understand the importance of art and permanence in civilized life. Maybe we will get to the point in the next millennium where Christopher Lasch's dream of art's primary source of being in the genial middle ground of everyday experience will come true. It's true and has been true for years in New York and L.A. and maybe it will also be true for Milwaukee, Akron, Dallas, Tokyo, and Belgrade.

Maybe I can find my own plot of genial middle ground and design something of northern Minnesota import. Frank Lloyd Wright founded a whole school of design from the genial middle ground of the midwestern prairie. The world ate it up—and continues to. Louis Armstrong found his in New Orleans, and the world ate that up, too. They both had their roots.

Why do we so often wait until our own funerals to go home? Surely there is some connection between staying put and a civilized life. How can anything grow without putting down roots?

15

Min og Bo

But I say, walk by the Spirit . . . the fruit of the Spirit is love, joy, peace, patience, kindness, goodness, faithfulness, gentleness, self-control. . . .

—GALATIANS 5:16–23

Busy talking on our car phones, having just dropped off the kids at day care, fortified by our twenty milligrams of breakfast Prozac, driving away from a suburban household chock-a-block with high technology (including global Internet connections and security systems), creeping on our ways to work (through a decaying inner-city neighborhood) to a stressful job in a devolving corporation located in a sterile industrial park or crime-infested city, under a smoky sky—many of us ask the following question: How civilized are we?

I find myself beginning an imaginary conversation with that question. I wonder what life was like in Florence during the Renaissance, or my mind drifts to Cotswold sheepherders or the well-ordered Shaker communities of the nineteenth century. I usually temper such fantasies with comforting thoughts like, life may have been less hectic in olden times, but I can't imagine having surgery without anesthetics, or sweating out a Florentine summer without air-conditioning. Progress is progress. We have to preserve what we've learned about civilization even as we rush to take advantage of the latest advance in technology. It's no religion, but it can—and should—serve a civilized vision of the world.

An industrial designer terminally preoccupied with the quality of life and human artifacts, maybe I'm merely overwrought and suffering from what designer Jay Doblin called "the curse of esthetics." Knowing that many other people in the modern world live as I do, I am perhaps wrong to be suspicious and dissatisfied.

We love nothing more than to examine ourselves and our lives, mostly our shortcomings and problems. We have invented social sciences and psychology and biogenetics and psychopharmacology. We have government-supported mental health research, a billion-dollar self-help industry, scores of psychobabble television talk shows, and a ponderous literature devoted to our social ills. (My father collected postage stamps depicting the history of our country. I collect magazine covers depicting the vast array of oddities and quirks our culture seems fascinated with. Maybe study has become a substitute for living.)

Gerald Rohlich, a natural science professor at the University of Wisconsin at Madison, was one of my mentors. He taught me to be a practical idealist, uncomfortable with merely defining problems through study and research. Now retired, he was an international authority in the 1920s on waterborne diseases. He lamented to me that so many dissertations were done on well water contamination in rural Wisconsin, before the state and academia finally got together to eliminate the threat of typhoid fever. Whole towns vanished, and hundreds died before something was done. Perhaps civilization has removed the bias toward action that characterizes what we think of as "uncivilized" cultures.

Yet for all the studies I read of AIDS, violence, community disintegration, the criminal justice system, I remain at a loss for any set of guidelines to live by, other than my inherited moral compass. Nor am I comforted by anthropologists' descriptions of noble savages of the past (idealized versions of reality that have been around for two centuries). Nor do I pine for the good old days of my great-grandparents. But I do think we have to act to preserve civility and civilization, contents and discontents. Civilization will not fend for itself.

Years ago, a few miles north of Copenhagen, near a restaurant in Solorud Kro, I came upon a church graveyard. In the dusk as I walked along a curving path, I realized this was like no burial ground I had ever seen. It's not that I'm a connoisseur of epitaphs or headstone architecture. Far from it. I dislike cemeteries for the most part. None seems a civilized place to visit, much less a site for all eternity—except for this one.

Perfectly placed behind a centuries-old church, but astride a much-used public pathway between villages was the most delightful cemetery imaginable. I was hardly sure it was a cemetery and not a park. It was more like a garden, complete with flower beds, trees, shrubbery, open and intimate places with wood benches to sit on and listen to the birds. More surprising still were the grave markers: no rectilinear stones at the head of three-by-six-foot plots, arranged in orderly rows, or flush to the ground for lawnmower efficiency. Each marker was different. Yet each was small and designed to be part of a carefully arranged gardenscape. A place to be tended by hands, not machines. The composition of boulders, plants, and annuals bespoke human care and respect.

One of the markers simply named a young couple killed during World War II. Carved into a small boulder were their first names, "Min og Bo" (Min and Bo), and the years they lived, 1921–1943. Somehow their tender age and the use of first names said to me that here lie two young lovers, or maybe a brother and sister. Other markers, while usually more informative, were also etched in natural materials. Sometimes small cast plaques like those used in urban gardens to mark flower varieties signified a grave.

Never had I felt so at home in a cemetery. The naturalness of the place lifted the pall of death, the memories of flamboyant funeral parlors, the dressed-to-kill corpses, the stench of sweet-smelling and dying cut flowers, the ubiquitous planters and the last monuments to status here on earth. This garden for the dead seemed like a natural hereafter on earth, a fitting rather than a utilitarian end to life. It remains an unforgettable place today.

Another cemetery comes to mind that also made dying seem less of an endplace to me. Near the village of Castletownsend, in County Cork, Ireland, I came upon a ragged, slightly overgrown cemetery of more conventional design. Some of the monuments were Stonehenge-like, worn by the ages. It was raining that day, and the wetness that turns Irish grass greener than any other made the dark stones glisten like diamonds.

Among graves dating back three hundred years was a fresh one yet to be adorned with a stone. It was marked by a touching display, yet by American standards bizarre. About a dozen neatly arranged, empty pint whiskey bottles circled a large and rather vulgar wreath of bright-colored artificial flowers covered by plastic. A roaring drunk, or a successful whiskey distiller, or simply someone who was dearly loved and sorely missed lay here. I'd like to believe the latter, and I found myself not thinking about

An Irish Wake

death but trying to imagine the toasts, the remembrances, and laughter all that booze must have elicited at the burial.

Maybe you couldn't care less about your final resting place. Most of us wind up in extensions of family plots, or in an urn over the fireplace. The lucky ones from rural or small towns will be buried in one of those pristine church lots in the country. The less lucky will come to rest next to suburban shopping malls or within earshot of airports and freeways.

Winston Churchill could have been buried in Westminster, or St. Paul's Cathedral, or some noble battlefield. No, he chose a modest churchyard in

Bladon, England, overlooking the Oxfordshire countryside and Blenheim, where Churchill was born.

In our times, we have removed death and the dead as far as we can from the living, as if we could avoid death ourselves by banishing the thought of it. We have immured it in hospitals, nursing homes, and funeral parlors. None of this makes death any less a part of life.

There can be civility in death. In our times, when living longer seems a certainty, can design help us live longer with dignity? Min og Bo reminded me that design—even of cemeteries—can intervene to preserve civilization and a civil life.

PART III

Paths of Civility

16

In a London Workingman's Pub

On a rain-soaked afternoon in 1976 while researching how lower-middle-class Londoners coped with housing, an English friend, an American businessman, and I ran out of gas in an industrial area of the city. A member of an English kind of AAA, the Englishman called for help and suggested we resort to a pub on the corner while we waited for gas.

The pub was modest, far from the gussied-up tourist pubs of central London but warm and cozy as only British pubs can be. A workingman's place. We took stools near the door at the U-shaped bar to watch for the service truck. The better part of an hour went by as the three of us sat at the bar with an assortment of tradesmen. Curious about British beer, I ordered one of those dark murky brews, warm to the taste and hardly thirst-quenching. I had only taken a couple of swallows when the petrol arrived. We got up to proceed with our mission. On my left, a complete stranger gently touched my arm and said, "If you are leaving, tell the bartender, otherwise he'll leave your drink here because he'll think you're coming back."

Even in American bars where I have known the bartender, much less in a new place, I have never felt such sovereignty over a drink or a place on a bar stool. To nurse a beer in a crowded bar in the U.S. is frowned on as a loss of revenue, and I've had unfinished drinks or barely finished meals whisked away in such a manner as to say, "Get on with it, customers are waiting."

I have never quite forgotten the feeling in that pub that as a complete stranger I was worth something in public. I later asked my English companion if this was simply a case of common courtesy or was I being politely reminded not to be so openly wasteful. He said that it's a matter of course that "a man and his beer and his place" are sacred even in the humblest of pubs.

I retold this story to

Clino, an Italian colleague who had just completed a physical exam at the Mayo Clinic. He complained about what he felt was rude treatment at the hands of American doctors and nurses in such a renowned clinic. He said, "I had to stand in line waiting for blood tests and carry my own health records. I felt like a cow in a slaughterhouse." (He did admit that the diagnosis was thorough.) Given the Mayo Clinic's reputation, he expected it to be more like a spa, and he anticipated much more deference and respect. Amused and hardly surprised, I said to him, "In America we have a saying: 'Who the hell do you think you are?' " He quickly retorted, "In Italy we have a saying: 'Do you know who I am?' "

I can't help believing that Americans' search for self-esteem within the boundaries of our splendid selves misses something more important. The great urge to look inward is blinding us to the world outside. Maybe it is the result of two centuries of hardy individualism and pioneer spirit. It makes little difference how perfect we may be—or how wealthy or how beautiful—when in public we can be reduced to nothing. Americans enjoy sovereignty in their homes, but in public we can and often do endure small and large measures of interpersonal indifference or outright alienation.

One of the marks of civility that springs immediately to mind is a public respect for everyone. Great leaders have this quality, a lesson I've learned from business leader Max DePree and his books. Everyone has a place in our world. Public environments (in addition to *pub*-lic houses) should allow us to feel some measure of self-esteem and belonging.

The kindness of strangers does have a role in maintaining our so-called self-esteem.

17

Walking Russell to School

There was a time when meadow, grove, and stream,
The earth, and every common sight,
To me did seem
Apparelled in celestial light,
The glory and freshness of a dream.
—WILLIAM WORDSWORTH

I went to Ann Arbor, Michigan, in the autumn of 1993 to work on this book and stayed in a delightful neighborhood south of the University of Michigan campus, a place of modest homes built before World War II. With large shade trees afire with fall colors, the scene during my stay was archetypical midwestern domesticity—a seeming heaven just west of turbulent Detroit. Academics and professional families live in what might be called reasonable prosperity. No one appears rich or poor here, but comfortable. By all standards this is a good place to live, rich in cultural choices and without the stress of bigger cities.

The houses in the neighborhood were never designed for the wealthy, nor were they meant to be start-up houses to grow out of. Some are two stories with generous windows, spacious but with no large rooms. Some are small and cottagelike. I'm delighted with the architectural diversity here, so unlike the dreadful monotony of the new developments.

Like the yard in my childhood house in St. Louis, backyards here are less important than front yards. With porches facing the street, families talk with neighbors, watch over errant children and pets. These houses present a convivial and distinctive face to the street, particularly at times like Halloween. I like to call these facial neighborhoods, places where one can read the life in the houses, not just pass by the empty elevations so typical of suburban neighborhoods. These houses have identities, not simply addresses. While some of the original owners still live here, most have moved away, and in all probability the new neighbors are better educated and better off. What matters more is the heart of this place. A common good transcends whoever lives here; this is a real neighborhood.

This good is defined not by electronic alarms, guard dogs, and high fences but by unassuming comfort achieved with reasonable means—neighbors looking after neighbors, within walking distance to bus stops, stores, the dentist, parks, playgrounds, shaded streets, and schools. Fortunately, walking here is integrated into life, not dictated by a specific jogging path.

I don't want to use the word architecture, because of the immediate associations with wealth. Few middle-class families live in architect-designed dwellings. Somehow in this neighborhood architects were unnecessary, for builders and home buyers understood that each house could be different.

Just a few blocks away on the edge of a lovely park surrounded by homes is the neighborhood public elementary school. The school was probably built at the same time as the houses and is an eclectic design with a kind of scholastic demeanor about it: a three-story, red brick building with hints of early American motifs, paned double-hung windows trimmed in white, and quarry-tiled floors. It doesn't look like a factory or corporate villa as do so many new schools today.

Russell, a young friend of mine, invited me into his first-grade classroom, I was transfixed by its similarity to my own first-grade room. Filled with colorful displays, maps, artwork, and projects, it was neither overly tidy nor hopelessly chaotic. Ah! I said to myself, everything here is as it should be. Civilization, at least as I remember it, is alive and well. Russell is a lucky young boy. He lives in a real neighborhood, connected to culture and nature. He's being educated in a good old American public school. An obviously good public school, where parents work regularly with dedicated teachers.

Is there anything wrong with this picture of civility? Unfortunately, yes. Even in this idyllic setting, there are hints of decay. The civilized enamel has broken down in spots. Without intervention, the people in Ann Arbor may lose the entire tooth. While Russell may be unaware of the fact, his life is closely watched. Even here there is little unsurveilled play and loitering to and from school. Russell's parents feel an unspoken need to surveil Russell's whereabouts for all the obvious reasons, fear that even within this nearly perfect setting should they leave him alone or unobserved for even a minute, he may come to harm. I doubt that Russell ever feels truly free, truly free by himself at such a tender age—free to run and hide, to play tag, to play catch, to appear unannounced at a friend's house, to pee behind a garage, to tread surreptitiously on a crabby neighbor's grass, to chalk-mark the sidewalk, to get good and dirty or soaking wet, to play after dark or simply to hang around.

On television during my visit to Ann Arbor, an elementary school principal in Philadelphia argued with the police chief about a poorly timed drug bust a block from her school. She asked, "Why do five patrol cars with sirens screeching have to speed past the school yard at dismissal time?" A kid's sense of well-being wilts amidst violence and chaos. Pumping up inner-city kids' self-esteem through classroom techniques accomplishes little when they witness shootings on the way home from school. No one can be completely free from harm. But without a minimal sense of public safety, life can be rubbish in America.

Isn't the point of these stories of Ann Arbor and Philadelphia this: In subtle and heavy-handed ways we condition ourselves and our children to accept a less-than-civil status quo. In an obvious way, the police chief in Philadelphia did not stop to ask the

effects of his actions on the civilization in his city. In a more subtle way, the good people of Ann Arbor have done the same thing. Too much resignation is dangerous.

A sip of water at a public fountain should be tasteless, the air we breathe should be pollution-free, and the paths we walk should be safe and open to all. All the details of life add up to a civilized life—or the lack of one.

18

D. J. and Dursu

I am inclined to think that it is one of the permanent tragedies of life that the finer quality doesn't prevail over the next less fine.

—ALFRED NORTH WHITEHEAD

And it is clear, we have gone too far with this sort of artificiality in the belief that there ought to be mass-produced substitutes for every aspect of human life.

—MARGARET MEAD, World Enough

Architect Hugh Hardy lamented to me a few years ago that the most disturbing characteristic of young graduate students applying for work was that they had more interest in designing a building than actually building it. This kind of thinking has given design a tinge of effete uselessness. Design is much more than talk or sketches or plans, which almost any person can engage in. Design, like music, offers up its true meaning and significance only in its performance.

The old chestnut argument between design-as-intention and design-as-execution won't be roasted here. It doesn't merely show up the difference between the natural and the artificial. Artificial roses today can fool even expert eyes and noses. What cannot be made artificially—at least not yet—is the birth, growth, and death of roses, the drama of mortality, which in the end gives meaning to earthly experience. Artificial roses pressed in a diary will never wither. The roses in my backyard certainly will.

Part of design's contribution to a civilized life is to help people make meaningful connections to the real world, not only the world as created by technology, but the world that existed before technology and that will exist after it. Let us not forget reality for the sake of appearances and convenience—lest the secret scandal uncivilize us for good.

Strangely, two quite different people come to mind who had the same moral sensitivity about connections and caretaking in the environment: D. J. DePree, a highly successful, humane, and sensitive American businessman and the founding father of Herman Miller, Inc., the company that produced the designs of George Nelson and Charles Eames; and a fictional character, a Mongolian tribesman from Kurosawa's epic film *Dursu Usala.*

Throughout D. J.'s life, he had the habit of cleaning restrooms after he used them, of leaving toilets and sinks spotless across the country. Unlike those well-intentioned groups that clean up trash on the highways, D. J. didn't take credit on a sign polluting the very thing supposedly cleaned up. D. J.'s caretaking in public restrooms was completely anonymous. No one knew of this charming habit until his son Max DePree mentioned it at D. J.'s funeral. Of all of his achievements in modern design and business, it was D. J.'s humble and anonymous caretaking that his son chose to remark upon. D. J. left things better than he found them. He also left people better than he found them.

In the course of guiding Russian surveyors across the frozen tundra of Mongolia in the early 1920s, Dursu, the protagonist in Kurosawa's film, tried in vain to teach basic survival to the entire regiment. After warning them of the dangers of Arctic storms and showing them how to make makeshift enclosures out of swamp grass, he witnessed their deaths when most of them ignored his advice. Those who listened survived and came to respect his wisdom on the rest of their frigid journey.

Like D. J., Dursu left things better than he found them, for he also had a habit of carefully cleaning his campsite before leaving of all debris, cutting and stocking a fresh supply of firewood for the next traveler. D. J. and Dursu had a deep and abiding understanding of our social and earthly interdependence. Through caretaking they expressed a civil relationship between themselves and the environment, between themselves and people they would never meet.

My wife, Sharon, likes to talk about my own eccentricities. Just before midnight while delayed in Denver, Colorado, on a cross-country Amtrak train a few years ago, she was shocked to see me in my pajamas in the dead of winter washing our compartment's filthy window from the station platform. I was determined to see the glory of the Rocky Mountains the next morning through clean windows. Even worse, I proceeded to clean the entire compartment with bar soap and tissue, for it was equally dirty. I'll admit this may be taking things too far, but I think we need a caretaking relationship more than that afforded us through being mere landlords, renters, travelers, tourists, or squatters. We need to become involved in the daily life of our world and the environments we create on it.

Imagine how different our cities would look if we cared for them as well as we do our homes and backyards. In preparation for Gorbachev's visit in 1991, Minneapolis turned out its young people and volunteers to clean up the freeways. They planted new trees and imaginative floral landscapes on all the routes the entourage would travel. The ugly, uncared-for residual space so common to our everyday urban commuting came alive with color and a sense of civic joy for the rest of the summer. We had cleaned up the house for company.

In St. Paul recently, neighbors—mostly renters—reduced crime by over fifty percent in an inner-city, three-block neighborhood. They began to act like owners. They got together and cleaned up all the trash, fixed up abandoned property, painting and building fences, so that crack dealers wouldn't have free rein over backyards and alleys. By becoming personally involved in their environment, they changed it—and they changed their lives.

Caretaking reverses the ubiquitous "parting psychology" associated with the decay of rusting factories, decaying neighborhoods, junked cars, trash, vacant lots, and littered lakes and rivers. Sadly, Disneyland may be our just reward for ignorance of history. Disneylands are unremittingly nostalgic extracts of a hoped-for utopia, an unreal architecture of sleek and functional and phony urban public transit, a remanufactured recollection of once-vibrant small towns, carefully attended to (hordes of Disneylanders clean up the place every night), safe, with clean streets, small shopfronts, and the friendly drugstore on the corner. I often ask, "If Disneyland is so great, why don't we live there and visit blight parks for amusement?" Or there is the infamous Mall of America in Minneapolis, Minnesota, the world's biggest shopping fix. All the fears of urban violence, filth, and decay are taken care of here for us. It is, in the mold of Disneyland, a sterilized city in and of itself, requiring no responsibility for caretaking. It is all too easy and, therefore, doomed to remain an unreal anomaly.

But these are all small potatoes compared to the larger questions: Where is the invisible caretaking hand that accepts the civic responsibility for clean water and air, good roads, bridges, a balanced transportation system? My neighbors and I would love to see downtown Minneapolis cleaned up every night. A massive injection of great public works on the scale of Ronald Reagan's Star Wars would make Americans feel somebody is caretaking the future for our children.

Certainly our collective self-worth would improve by proclaiming that there is virtue implicit in heroic efforts being made to take care of culture and its environments. "Housekeeping" used to be—and still is in some places—a respected life goal. To "keep a good house" was a compliment, not an expression of pity for someone without alternatives. I suggest we become a *nation* of housekeepers.

A caretaking relationship is necessary for us to survive, much less prosper. Nothing will change until we take personal responsibility for caretaking of not just what we own, but what we all share.

19

Lace Curtains in the Police Station

I do not love thee, Dr. Fell.
The reason why, I cannot tell,
But this alone I know full well,
I do not love thee, Dr. Fell.

—THOMAS BROWNE

Polarization results from drawing distinctions. In its worst form it divides races, states, and nations. It separates black from white, rich from poor, beautiful from ugly, cheap from luxurious. What's hot from what's not. We are fascinated with pop dichotomies that neatly divide our world into arbitrary categories, and we are tempted to take this easy way out, forgetting about meaningful distinctions like historical and ethnic roots. In America the need to draw distinctions is particularly acute in commercial enterprise. While meaningful distinctions are important to self-definition, historical and ethnic roots, particular places and diverse climates, they go unremarked by those determined to control market behavior.

Corporations in the name of "economies of scale" insist on serving up the same stew to a targeted audience. Advertisers live in the world of false dysjunctions, creating and exploiting artificial distinctions between the young and old, the lower, middle, and upper classes, the mobile and immobile, people of color and colorless people, the rural and the urban, the urban and the suburban, the educated and the uneducated, and most of all the rich and the poor. Commercial zeal in drawing distinctions has influenced education and politics today in the pursuit of correctness. I can't say enough about the perverse effects of phony correctness on esthetics and design. The American-Puritanical mania for perfectionism and correctness mutes the individual freedom to make meaningful esthetic choices. It reduces our capacity to empathize with places, things, and experiences rich in contradiction, originality, and eccentricity. It market-manages imaginative design and entertainment into artificial group recipes for mush in community building, commercial and domestic architecture, cars and kitchens of a future that will never be. The desire to be correct at the expense of true expression relegates risks and flights of fancy to the dustbin of facile polarizations. Art critic Robert Hughes claims that political correctness "divides the sprawling republic of art neatly into goodies and baddies. . . ." Civil society, it seems to me, pleases us all precisely because it recognizes that life mixes the good with the bad, the inspirational and the absurd, the chaotic and the tranquil.

"The peculiar beauty of good design is that, at its best, it mediates extremes without destroying them," says my friend Clark Malcolm. I also believe that design—more than almost any discipline I know—gives us grounds for hope and civility. That goes for the design of objects, places, and lives. Places have a quality of demeanor, tone, or deportment that can enlist a sense of joy, comfort, community or fear, dourness, and despair. Or all of these at once. Detroit and Disneyland easily contrast with each other, the poles of an insidious dichotomy. Most places mix joys and sorrows. Perhaps it's the balance of these demeanors that makes life both real and possible.

Just under the modern, glossy surface of the city and in the suburbs and malls of Minneapolis and St. Paul lurks an increasing sense of civil dysfunction. On an abstract level it can be seen not in walls of graffiti but in such things as newly built day care centers with the same architectural demeanor as police stations, or suburban high schools that look like mundane office buildings, or worse, prisons. Police patrol cars, sporting shotguns and caged passenger compartments, have the demeanor of Hollywood crash-and-burn violence, with no civic service role implied other than criminal apprehension. The police, dressed like mall guards, are festooned with the communication and assault gear of paramilitary forces, or lamely wear baseball caps, as if to present a softer image.

The stationing of police in local public schools and the soaring enrollment in private schools are evidence of the fear and anxiety of parents. Security and armed response signs sprout in every front yard where the service can be afforded. Gated communities and hotels like the Hyatt near the new Disney town called Celebration have the demeanor of plush, domestic ghettos or prisons, complete with guarded entries, something I've never seen in Europe or Japan. All of this, amid wealth and the trappings of normality, hints at an uncivilized public spirit.

While walking in front of the centuries-old courthouse in Basel, Switzerland (a city comparable to Minneapolis), in the central town square, I was dumbstruck at finding that the main police station had a quaint, arched doorway, a door with wrought iron, heart-shaped decoration, and lace curtains in the windows. Lace curtains in the police department. Wow! In Schaffhausen, a smaller city of

90,000, upstream on the Rhine, I found another police station and city jail with more lace curtains, wall murals, and even geranium plants in the barred windows.

I waved at one of the prisoners, and he waved back. I thought, If I were in prison, here is where I'd want to be. How can a prison be so domestic-looking, nothing like the grim utilitarian or menacing demeanor typical of American jails. Most of all, here I sensed an irony about punishment unknown to Americans. These prisoners were smack dab in the center of town overlooking beautiful scenery and architecture and a vibrant city life complete with outdoor cafes. Certainly the worst punishment for me would be to be behind bars looking out on the bustle of everyday life, not lobotomized in some far-flung rural campus flogging a rock pile. Somehow here I sensed a form of civil wisdom at work. A reminder to the chained and unchained that there is an alternative to one's modus vivendi. By making the goodness of life vivid, some deeper meaning can be reached about human error. In the United States,

we insist on punishment that reorders life in harsh prison terms, that supposedly rectifies human character through isolation and estrangement from normal life. It doesn't work. We have more people in prison per capita than any developed nation.

We could change the demeanor of law enforcement. Imagine small-scale local prisons in a domestic architectural style erected in towns and neighborhoods, like schools and hospitals, where families of prisoners can get together, instead of mega-prisons built in remote archipelagos. A friend of mine works with alcoholics at a regional state penitentiary and finds the prisoners very much concerned not just about "getting out" but about their loss of known local experiences, especially those connected to families. The irony and pain of being imprisoned in one's own backyard might create a less apathetic civic consciousness on the part of prisoners, prison guards, and the community at large. Who knows, maybe a kid's ability to wave at his mom or dad through the prison window might create the tension required to build a deeper empathy for the millions living in America's prisons.

Or consider adding minimum-security prison wings to public schools in major cities. The irony of being part of life but being barred from it could work its way with youthful offenders, to say nothing of maintaining a more intimate and integrated network to their friends, families, and textbooks. Public shaming may seem like a Puritan throwback, but in a society hell-bent on self-image, it may be a civil alternative to the invisible woodshed.

Drawing distinctions and erecting barriers are two different things. We cannot afford to put a forbidding face on public servants and thus isolate them. The civilized mixture of differing human impulses—lace curtains in a police station—somehow reminds us that pulling together is better than pulling apart.

20

A Scottish Way of Retirement

I'm sixty-three and I guess that puts me with the geriatrics, but if there were fifteen months in every year, I'd be only forty-three.

—JAMES THURBER

Will you still need me, will you still feed me, when I'm sixty-four?

—THE BEATLES

In the United States, environmentalists and my design colleagues are very much concerned with the wisdom of the sustainability movement. Unfortunately, concern for recycling and conserving life-sustaining resources, plant and animal species, soils, forests, minerals, and cultural legacies rarely or directly focuses itself on preserving a civil way of life—a full life, like a good story, with a beginning, a middle, and an end. One of these days, my body will recycle itself, but I need some help with the sustaining part.

Having just passed sixty, I am regularly confronted with unwelcome hints that my life is waning, that I'm into the last innings of the game, that the best of life is but a memory, and that I should get about the planning for my retirement. In short, I should be developing a plan and designing my forthcoming demise. My best friend humorously reminds me that males reaching sixty have but twelve good years left to live. A terminal vision, no doubt based on actual insurance statistics, and in my case—males in my family rarely reach seventy—somewhat ominous. I'd rather be comforted by recent findings that humans can live 120 years, aided and abetted by the life-extending miracles of modern medicine. The pace of my life now reduces hours to minutes, days to hours, and months to weeks. I take life a month at a time. I may be living longer, but I'm certainly living faster. The fact that we are living longer than our recent forebears is no solace at all, for the question is, Are we living any better than they did?

I feel the pressure on all members of society to invest in longer periods of post-high school and college education, to sustain career-long learning, all the while preparing to avoid the position of an economic liability past the age of fifty-five. With the boomer hordes about to experience what amounts to an enormous, last-third-of-life vacuum, the inevitable devolution of Social Security and health care benefits, and the expected increase in life span, retirement as we know it is about as viable as the Cleaver family in the cyberworld. The promise of assured leisure in older age is gone. The specter of working forever involuntarily is a bad idea whose time has come.

Slightly older than the boomers but part of their social caste, I ignore the appeals for deals from AARP, but I can't ignore store clerks offering me discount cards at the cash register, nor can I evade questions from friends who ask me why I don't slow down, or why I'm embarking on a rigorous reading program of the classics, or why at my age I am building a new house. Every evening just before dinner, June Allyson's adult diaper commercials on TV assure me that I needn't worry about incontinence. I do pee more frequently than I used to, but June as diaper pitchperson is hard to take. I held her up as an object of desire in the late 1950s. Marvin Minsky, the esteemed mathematician, while speaking on the future of cyberspace, told about a close friend who died on the handball court at my age. He had ignored Minsky's theory that humans have only a finite number of steps in their bodies. Minsky had warned his friend not to use them up foolishly. I am aware of the slight weariness I feel deep in my bones. I sense the need to ration any steps I have left.

Without question, demography has its uses, but as expressed in the American fetish for market-driven stereotyping, it hovers over generations and locks them into short little decades of social viability. Think of the flappers in the 1930s and the rappers in the 1990s and all the DINKs, beatniks, boomers, hippies, yuppies, and Generation Xers. We are forever marked and market-segmented into tidy time warps, like it or not. Women, more so than men, have but a few years to be considered youthful and beautiful. Many women past forty and increasing numbers of men feel compelled to freeze their flesh in time either through surgery or exercise or makeup, fighting a hopeless battle with age for the rest of their lives.

America now has its ghettos of silver hairs and, worse, its ghettos of the elderly devoid of any mixed-age daily activity. Living in what amounts to planned stages of death hardly supports the idea of sustainability. It smacks of planned obsolescence. Like sour cream or batteries, we have a limited shelf life, and once all the juice is squeezed out we feel pressured to pass unobtrusively into oblivion, with or without the help of the good Dr. Kevorkian.

Someone told me that many artists live to a ripe old age because they are fundamentally performance-oriented, because they live from project to project, and the stress of projects sustains their need to learn, create, and perform. Certainly, the late, productive years of Henri Matisse, Pablo Casals, and Frank Lloyd Wright come to mind. Wright began what architectural historians have recognized as a major creative period after sixty-five. I can't imagine seeing Wright at leisure on a yacht or the golf course. But then, these were exceptional and highly talented people. What about us working stiffs, those of us who for one reason or another have had our spirits broken and lives truncated by repetitive work, normal health, dislocation, or misfortune? What about the mass of us whom Thoreau described as living lives of quiet desperation? What's to keep us going after we've been put out to pasture?

Certainly the political sensitivity of even the most hard-nosed conservatives in Congress regarding the potential disintegration of the Social Security system suggests that great fears exist in the U.S. for those entering retirement. These fears transcend race and gender. The chance of being poor and disconnected to the mainstream of American life is harsh and uncivilized.

It seems like a good time to imagine more options for retirement, to make possible a more civilized foyer for those of us waiting to enter the Final Cinema. For people like myself, who love what we do and see little distinction between work and leisure, there is always hope. But for others, perhaps a short story about a Scottish way of retirement might prove worthwhile, a story about a seventy-year-old married couple in Scotland practicing a delightful, unretired retirement.

A few years ago, my wife and I were exploring the rugged, northern coast of the Isle of Skye and came upon a sign which read WOOLEN KNITTED SOCKS FOR SALE. A charming, white stone cottage and small farm building drew us down a winding road through a lavender-colored heather field overlooking the whitecaps of the Minch. We knocked on the cottage door and were fortunate enough to arrive at tea time. Seemingly glad to see us, the friendly couple insisted we drink their tea before we looked at their merchandise. In no hurry, we sat in the small parlor which smelled of decades of peat fires, hav-

ing tea, eating delicious scones slathered with clotted cream and homemade raspberry jam.

It remains quite amazing to me how easily and with what conviviality the Scottish folk engage in conversation with complete strangers. I have a neighbor who visits Scotland regularly simply to learn the art of conversation. Though we came to buy socks, we talked for more than two hours about Scotland, America, World War II, grandchildren, growing old, and the beautiful scenery around their house. Most interesting to us was why and how this particular couple regarded what they were doing as *retirement* and why they started their small but economically viable business knitting socks for local markets. He was a retired British Army officer, and she was a self-described gardener and part-time artist. They had decided early on that they both wanted to avoid the typical work-no-more oblivion or the equally unappetizing alternative of the dole and a restricted army pension. While they did take advantage of free state health care, they wanted to be involved in meaningful, daily work, particularly during the harsh, Scottish coastal winters.

I sensed a level of pride in their purposely designed last third of life, and things were working out as they had planned. They had decided on knitting, though they were untrained in the craft and the technology. They sold socks in local towns but worked no more than four hours a day and no more than three days a week. In addition, they raised raspberries and Australian shepherds.

After tea, we walked into a small, barnlike structure, white-washed inside and out, and with a flat, stone floor. Fully expecting the typical gift shop atmosphere with knickknacks, perfumey candles, and candy, we were surprised to find a tidy little knitting factory, complete with sophisticated, but used, Toyota knitting machines capable of significant levels of production. The walls were lined with spools of colorful wool and wool-cotton blends, and a single sewing machine that sewed up the hole left in the toe-end of socks sat in a corner. No attempt was made at *marketing* their wares; most if not all of the socks were in cardboard cartons ready for distribution. The socks were beautifully made, rich in heatherlike blended colors, cashmere soft, and oblivious to past, current, or future fashion—simply very nice socks, made of good materials, extremely comfortable (I'm still wearing them after three years), and above all reasonably priced.

While they sold a few pairs to tourists like us, their main market was small clothing stores in local villages and towns like Portree. In an aging and modest little English car, they delivered their weekly inventories, which sometimes reached two hundred pairs. They told how much they enjoyed working with and befriending the local merchants and customers. They were fulfilling the dream offered by Christopher Lasch: Art or "enterprise" finds its source in the genial middle ground of everyday activity.

What a small but entirely memorable joy to know that the socks I'm wearing came from someone I knew! The labor they incurred was honorable, the work creative and rewarding, and they weren't sweated over in a far-flung slum. It seems the same could be true for shoes, clothing, furniture, tableware, linens, foodstuffs, and the rest. Our franchise-blighted market economy needs to be deconstructed and become more responsive to small enterprise. Why not? This kind of enterprise would free up the shackles on design. To repeat, design, after all, as Ralph Caplan says, "is a local issue and opportunity."

I asked the couple if they were peculiar in their aversion to leaning on the local or federal government in retirement. They said no. They claimed their way of life was common in Scotland. As Sharon and I left, we talked at length about what we sensed as a vital way for people to live in retirement. We admired their gusto, their comfort in a reasonable level of prosperity, their real involvement in the community and local economy. We were so impressed that we set about planning a similarly civilized way of retirement for ourselves back home.

Designing a way of growing old may be more difficult than designing a car or a house, but it certainly pays larger dividends. Retirement too often becomes the failure of a dream rather than a chance to renew our vitality. Growing old should be yet another chance to take charge of our reality and civilize it.

21

Past with Present

The British government led by Clinton-clone Tony Blair has decided to redesign the venerable and familiar red telephone booths in Britain. Too old-fashioned, too un-technical, too redolent of party lines and actual human operators. And so my biggest design bugaboo raises its head: the sad belief that adopting the style or trappings of what the world currently calls "cutting edge" requires us to trash a tired and befouled past. It's like remaking Cambridge in the image of Palo Alto. It's an assumption that a new look will empower social or economic revolutions. Singapore, which now looks like Houston, comes to mind.

I have spoken a great deal about the past in this book, both its civility and its inconvenience. I've also dealt with the present, its incivility and its achievements—technical and otherwise. Design should preserve the best of both—mediate extremes without destroying them—and thus build a new civility. Designers and design could have arrived at a much more civil solution to the British telephone booth dilemma.

The classic British telephone booth is being replaced with a super glass and stainless steel, transparent, high-tech (read Americanized), red plastic-roofed, modular design. Soon to be gone is one of the most charming and civilized pieces of urban architecture in the world. Like the ones in the United States, the new booth looks like a glass coffin. Gone is the sense of small, cozy, semiprivate architecture. Gone is the combination of playhouse and official business. Shortly, I'm sure, the red English post boxes will give way to some stainless steel, ATM-like device.

A new design could have mediated the tensions and extremes of techno-coolness and a critical piece of British history. The old standby telephone booth could have been kept as a brightly colored piece of street architecture, giving a sense of privacy through small-paned glazing, yet improved with better lighting, space for the disabled, better ventilation, adjustable digital interfaces, voice recognition, and all the rest of the promising stuff of global communications. Neither the British—nor we—have to trash the past to take advantage of the techno-conveniences of the present.

The essence of civility expresses both past and present. Let me give you another example. As my friend Clark Malcolm and I were walking to a meeting one bright November morning in Manhattan, I tripped on a manhole cover and broke my hip. I'm not exactly the model of svelte middle age, but I'd never thought of myself as clumsy. Yet there I was, felled by a piece of steel protruding from the surface of Fifth Avenue.

The manhole cover may have been the height of uncivility, but random New Yorkers leaped to help me. Some stopped traffic. Some helped Clark get me out of the street. Others volunteered phones to call the paramedics, who arrived at the scene minutes later. Then we were, Clark and I, swallowed by the unfamiliar health care system in New York. I was x-rayed, poked and prodded, grateful for the technology and the kindness of strangers.

I was amazed at the present state of medicine that could supply me with a new hip. Robert Priester, one of the doctors I got to know, told me that not so long ago an accident like this would probably have been a "terminal event," as he so politely put it. As it happened, I got my new hip the next day, was walking slowly the day after that, and in a week was back home in Minneapolis.

I tell you this story not to convince you of my bad luck, but to let you in on a secret to civility: the best of present technology combined with simple, old-fashioned understanding of our own humanity. One without the other would be uncivilized. Kindness without my new, titanium hip joint or Demerol would have left me just another terminal event. All the titanium and painkillers in the world and no kindness would have somehow diminished me as a civilized human being—and diminished those around me. And I certainly would not be willing to return to Fifth Avenue. The kindness kept my *spirit* from becoming a terminal event.

Surely there are ways to improve the bodies of British telephone booths without bruising their spirit. In our cyber-rush to the future, let none of us forget the civility of the past—and the present.

Epilogue:

Nobody Asks

"Oh, my God!" Ignatius mumbled to himself, watching the silhouette of the streetcar forming a half-block away. "What vicious trick is Fortuna playing on me now?"

—J. K. O'TOOLE, A Confederacy of Dunces

Now I have wandered from topic to topic in this book, from observation to fulmination to suggestion. The problem of creating a civilized life is not an easy one. I consider it a triumph when I can carve out another niche of civilized design or behavior in my own small corner of the world, Minneapolis. Yet many people are trying to do the same in *their* small corners of the world. They are identifying and investing in their neighborhoods. They are heating their sidewalks in winter, and somewhere, I hope, they are measuring the windows of the local police station for lace curtains. They are asking, Why do things have to be the way they are? What can I do to civilize myself and my community?

As a designer, I have struggled with problems ranging from humanity and efficiency in offices to compassionate living environments for older people. I know the enormous difficulties imposed on us by the vast and unwieldy corporate-government-cultural complex. It isn't easy to make a change. But I have a sneaking suspicion that too few of us are trying.

Like many people, I'm annoyed by the contradictions in the design of modern life: mobile homes that move only in tornadoes; office furniture designed to be flexible but shifted only from the office to the Dumpster; suburban homes designed for resale but not for living. I realize that the contradictions and complexities won't go away, that I have my own contradictions, that there are no simple answers, that design itself never seems to stay in one place with a comfortable definition for all.

Civility, for me, is like that. It is a feeling, an attitude, a state of mind. In my darker moments, a civilized life seems like a complete fantasy. But most of the time, the idea of living a life civilized by design and designers is a dream, parts of which have come true in different places around the world and parts of which will come true any minute in my own backyard.

I hope this book reflects the way many people and many designers see the world. At least it presents my perceptions and a view that at its core is optimistic about the future. For better or worse, most of my design has been future think and do. Granted, my search for civility, for comfort, for essences rather than attributes in nature, technology, and social life, may be eclipsed by the potential of a digital future. At the end of the twentieth century, the ability to make judgments, even by artists about authenticities, has been clouded by the convergence of the real and the unreal and the emerging virtually real. I remain unfazed in my belief that it's fundamental to human nature to be critically conscious of whatever part of the reality spectrum one finds vital to life.

My plea here is for more of us to become designers, or at least to think more often like designers. It really is a matter of resolve—of involvement. It's a matter of concluding that we can make our lives more meaningful without charging it to Visa or American Express. We can connect our lives and those of our children to natural realities, health and sickness, ability and disability, kitchens and bathrooms, slaughterhouses and grocery stores. We can alter the ways we work so that we get a glimmer of humor and playfulness even while sitting on the Santa Monica Freeway during rush hour. We can increase our capacity for life by closely scrutinizing the objects around us and changing them. We can profess to ourselves and our friends that there is universal truth in beauty and transcending beauty in truth.

Naturally, I think designers can lead the quest for a civilized life. And I think they should. If nothing else, it will save us from spending our lives dressing up gadgets invented by marketing experts with sales goals. Design is still not the oldest profession, no matter how many characteristics it shares with the oldest. But why can't design come to be known as the most pervasive profession?

SELECTED BIBLIOGRAPHY

Alexander, Christopher. *The Timeless Way of Building.* New York: Oxford University Press, 1979.

Anderson, Walter Truett. *Reality: Isn't What It Used to Be.* New York: Harper Collins, 1990.

Bach, Richard. *Biplane.* New York: Dell Publishing, 1996.

Barzun, Jacques. "The Paradoxes of Creativity" in *The American Scholar.* Summer, 1989.

Becker, Marvin B. *Civility and Society in Western Europe 1300–1600.* Bloomington, IN: Indiana University Press, 1988.

Berman, Marshall. *All That Is Solid Melts into Thin Air.* New York: Simon and Schuster, 1982.

Braudel, Fernand. *The Structures of Everyday Life: Vol. 1.* New York: Harper & Row, 1981.

Burkhardt, Jacob. *The Civilization of the Renaissance in Italy.* New York: Penguin Books, 1990.

Caplan, Ralph. *By Design.* New York: St. Martin's Press, 1982.

Drexler, K. Eric. *Engines of Creation: The Coming Era of Nanotechnology.* New York: Anchor Books, 1986.

Elias, Norbert. *The History of Manners.* Oxford, England: Basil Blackwell Ltd., 1978.

———. *The Civilizing Process.* Cambridge, Mass.: Blackwell Publishers, 1994.

Emmeche, Claus. *The Garden in the Machine: The Emerging Science of Artificial Life.* Princeton, NJ: Princeton University Press, 1994.

Florman, Samuel. *The Civilized Engineer.* New York: St. Martin's Press, 1987.

Garbarino, James. *Toward A Sustainable Society.* Chicago: Noble Press, 1992.

Grudin, Robert. *The Grace of Great Things: Creativity and Innovation.* New York: Ticknor & Fields, 1990.

Gruchow, Paul. *The Necessity of Empty Places.* New York: St. Martin's Press, 1988.

Hiss, Tony. *The Experience of Place.* New York: Knopf, 1990.

Hughes, Robert. *Culture of Complaint: The Fraying of America.* New York: Oxford University Press, 1993.

Ingleton, Roy. *Arming the British Police: The Great Debate.* Essex, England: Frank Cass & Co. Ltd., 1997.

Lasch, Christopher. *Haven in a Heartless World: The Family Besieged.* New York: Basic Books, 1977.

———. *The Minimal Self: Psychic Survival in Troubled Times.* New York: Norton, 1984.

———. *The True and Only Heaven.* New York: Norton, 1991.

Miller, Daniel, ed. *Acknowledging Consumption: A Review of New Studies.* New York: Routledge, 1991.

Milne, A.A. *Not That It Matters.* New York: E.P. Dutton, 1920.

Nelson, George. *How To See.* New York: Whitney Library, 1958.

Nussbaum, Martha C. *The Fragility of Goodness: Luck and Ethics in Greek Tragedy and Philosophy.* Cambridge: Cambridge University Press, 1986.

Sennett, Richard. *The Fall of Public Man.* New York: Norton, 1986.

Schumacher, E.F. *Small Is Beautiful.* New York: Harper Row, 1973.

ABOUT THE AUTHOR

A celebrated champion of ergonomic office furniture, Bill Stumpf has taught and practiced industrial design for almost forty years. His chairs have won many awards around the world. Stumpf lives in Minneapolis with his wife, Sharon.